CHARKHA AND CHIP

CHARKHA AND CHIP

Rural Industrialisation and Technology

Editors

Kamal Nayan Kabra

Laxmi Dass

GYAN PUBLISHING HOUSE

New Delhi–110002

Charkha and Chip: Rural Industrialisation and Technology

Rs. 540

ISBN: 81-212-0883-1

Published in 2006 in India by
Gyan Publishing House,
5, Ansari Road, Daryaganj,
Delhi-110002
Phones: (011) 23282060, 23261060,
Fax: (011) 23285914
E-mail: Gyanbook@vsnl.com

Laser Typesetting by: Tact Computer Graphics, Delhi
Printed at: Salasar Imaging Systems, Delhi

Contents

Preface

Modern industrialization—a prime defining feature of the present era—is generally taken as synonymous with the preponderance of large scale, mass production enterprises in an economy leading to enormous increase in labour productivity, urbanization and a large and growing middle-income group specialized in the generation and utilisation of high technology. Generally it is this section of society and the entrepreneurs who are enabled, by and large, to consume the goods produces by the modern industries as the actual working classes and those who are not incorporated in the process of industrial development are able to join the process of consuming the goods produced by the modern industries only marginally. (The quantitative gigantism, an intrinsic characteristic of industrialisation and reflected in the rates of growth of value-added, ignores, nay, practically runs down the significance of the qualitative issues like the appropriateness of the product-mix, technology, energy-use pattern, organizational and management forms, location of production facilities, life styles, values, sharing the consumption boom, indigenous cultural variables and parameters and social relations associated with the industrial enterprises, i.e., the pattern variables associated with industrialization). According to the popular, mainstream theories of industrialization the small industrial enterprises, largely based on traditional technology, turning out traditional products and located in a fairly dispersed manner, gradually give way to the superior competitive strength of the large, high tech industrial units spawned by the modern scientific-technological revolution. (Contrary to this kind of stipulation, one finds tenacious survival of small enterprises, household enterprises, own-account, informal enterprises, both in traditional and new lines of production, in practically all countries irrespective of the stage of industrialization). In fact, in many countries many such enterprises continue to flourish, new ones emerge and by adapting themselves to the new conditions come to occupy a significant, crucial role. (These facts speak volumes about the significance, relevance and bright prospects of these small new and old industries, including those located in the rural areas and small towns acting as the central place for a cluster of villages). Many factors, like the 'end of work', i.e., a sharp and steady reduction of employment opportunities inherent in the massive and continuous capital accumulation and technological progress, and associated social, economic, environmental and international imbalances, argue a case against investing all one's hopes in large, technologically sophisticated and mass production—based industries. These lessons of the experience with modern industrialization are particularly pointedly relevant for poor, high population-density countries with limited capacity to find investible resources within the framework of the existing system like India, with an overwhelming weight of farm employment in the workforce, who joined the race to industrialise in terms of their own national resolve at a much later stage in comparative historical terms. The relevance of this experience is particularly critical owing to her iniquitous integration in the international division of labour in utter disregard of her own needs, potential and aspirations during the long, dark spell of colonial domination. Unfortunately with the political independence, not all the elements of the deep-seated negative legacy have disappeared from the scene. (The view that without recreating in our midst an industrialization of the kind which led to the Western domination of the world economy, the countries like India have no worthwhile future to look forward to is among several such misconceptions). (There seems to be inadequate realization that instead of imitating the industrial growth path and pattern of the West, a lot has to be done to mitigate the persistent adverse socio-economic effects of the early industrialization of a few select countries in order to enable the countries like India to really prosper). This can be done by India by adopting and following an independent and appropriate path and pattern of development, including that concerning the industrial sector responding to the needs particularly of the poor rural sector.

It is in this context that a view has come to find increasing critical acceptance that the high cost and dim prospects of colonial, postcolonial mimetic and inappropriate industrialization can be reduced, and in fact, can be turned, over the course of time, into a source of positive and desirable pattern of industrial growth if instead of going in for a large scale and size of capital investment per industrial entity and each unit of output flow in every branch of production, the attention of policies and social interventions is directed towards the qualitative aspects of industrial expansion in terms of the suitability of the product-mix, technology, location, size of each production unit, ownership and management pattern, energy-use, etc. However, it must be clearly recognized that in none of such exercises it is possible, let alone desirable and helpful, to tie oneself up with the old, archaic and traditional technologies and products. On the contrary, one has to make specific efforts in terms of adaptation, upgradation and finding out of altogether new technologies in all the different senses in which pioneer thinkers like Schumpeter have expounded the concept of innovations. Clearly, such new approaches have to proceed in terms of more effective methods of societal participation and organization consistent with the prevailing patterns and expectations regarding the incentives mechanisms which can deliver. In other words, the technological improvement and revitalisation of the old and newer activities has to be understood in a fairly comprehensive sense and on a wide canvas. For the new situation and challenges facing these late-industrialising countries, well-designed specific RD efforts are called for rather than just relying on the existing stock of patented technologies which happen, by and large, to be the property of the mega MNCs of the Western countries who part with them on rather difficult and costly terms, without, of course, any appropriate adaptation. Obviously the small entrepreneurs and artisans are in no position to organize such activities by themselves, though inputs from them have to form important building blocks for the RD directed towards designing industrial activities appropriate to the local conditions under a benign guidance and sponsorship of a socially-oriented rather than market governed state.

The above propositions are all the more germane to the conditions prevalent in rural India and for their transition from the predominance of subsistence agriculture to a vibrant, people-friendly, diversified and prosperous economy, making creative use of the scientific and technological advances. The main thrust of the strategy required for this purpose may be summed up in a stylized form in a few propositions. It does not seem possible to industrialise India in a brand-based, non-involutionary manner which incorporates and benefits the common citizen as active participant in the process of industrialization and at the same time transforms agriculture and allied activities technically and organizationally without industrializing rural India in a big and decisive way. It means the economic activities in rural India have to increasingly make use of the tools, equipments and machinery incorporating scientific advances for responding to the needs of the production process and for turning out the consumption requirements, while easing and humanising the labour process and relations, extending and deepening the social and technological division of labour and for preserving and enriching ecology. But in order to bring about such an industrial transformation of the rural economy, it is imperative that, instead of imitating the industrialization experience witnessed so far, the industries to be encouraged are ruralised, that is, made responsive to and consistent with the specific needs and urges of the rural communities with non-exclusionary outcomes. Obviously, such a task would necessitate a purposeful, fine-tuned and prioritized set of scientific and technological innovations and appropriate socio-economic policies and political-administrative interventions. It also goes without saying that a really lot of work has to be done in order to be able to operationalise such a perspective.

While keeping these kind of considerations uppermost in our thinking, a number of papers which seem likely to make a modest advance in generating, spreading, strengthening and concretising widespread understanding over such issues have been put together in this volume. It is hoped that this sensitive and significant issue would receive ever more attention in the days ahead.

I must thank Shri Laxmidassji for associating me with this work and Shri Amit Garg for agreeing readily to bring it out. It may be mentioned that in the bulk of economic policy and development debates issues of this nature are only now beginning to surface. They redefine the terms and framework of the discourse. It is a really challenging task to be able to come to grips with the entire body of connected issues. This collection makes no such calims to be comprehensive and definitive. Our hope is that we have tried to put such issues upront and with some concrete lines of action. If we succeed in this modest task our purpose is served.

Kamal Nayan Kabra

Acknowledgements

Whenever one hears cottage industries, a casual impression is prevalent that people engaged in the advocacy that they are against the modernization or are opposing the important role of Science, which actually is not true. Keeping this stigma in view a conscious decision was taken to organize a National Seminar on the importance of "Technology to the Rural Industries". Top scientists and technology experts were invited to discuss the technological aspect for the growth of cottage industries. The technical session was presided over by not less than a person, Dr. R. Chidambram Principal Scientific Advisor to the Govt. of India. Important Scientists and Technical experts were requested to present their papers on different technological aspects of Rural Industries. This book is a result of those efforts and I am grateful to the scholars who could spare some of their valuable time to write these papers. There were some gaps. On some issues, the related papers could not be presented, those are also included in this book.

We wanted to bring out this book long back. But because of one reason or the other we could not do the same. Anyhow, we are glad that now this book is in your hands. We are aware that without the active co-operation and rich experience of the experts reflected on the papers included in this book, this book could not have been an impressive one. The efforts and time devoted by the experts to write papers on different aspects of appropriate technologies for the rural industries has been a commendable job. We acknowledge the contribution of these experts. We are also grateful to Dr. R. Chidambram for presiding over the technology session of the seminar and steer the discussion in tune with the subject.

We are quite hopeful that this book shall remove the stigma on the activists of cottage industries that they are anti-technology or anti-science or anti-RD. We are also thankful to Khadi and Village Industries Commission for giving financial assistance to organize this important seminar.

The seminar was organized under the banner of Akhil Bhartiya Gramudyog Mahasangh. ABGM is a national level confederation of all federations working for the village industries. The main objective of ABGM is to create a congenial atmosphere, where the cottage industries can grow and branch out. We are also thankful to all the members of ABGM for extending full support and co-operation to perform our responsibilities.

Laxmi Dass

List of Contributors

1. **Kaml Nayan Kabra,** Former Professor of Economics, Indian Institute of Public Administration, New Delhi.
2. **Laxmi Das,** Former Chairman, Khadi and Village Industries Commission.
3. **D.R. Gadgil, Late Professor Gadgil** was Deputy Chairman, Planning Commission, Government of India, Vice-Chancellor, Pune University and Founder Director, Gokhale Institute of Economic and Politics, Pune.
4. **Y. S. Rajan,** Former Executive Director, Former Vice-Chancellor Punjab Technical University, Punjab.
5. **S. Chatterjee,** Adviser, Office of the Principal Scientific Adviser, GOI, Sunil, K. Agarwal, Scientist, Science and Society Division, Department of Science and Technolgy, GOI.
6. **Ram Prasad,** Former Scientist, CSIR and System Director, TNDRD Technology Nursing and Digital Research and Development, Delhi.
7. **A.K. Bhatia**, Former, Adviser, KVIC.
8. **Dr. Rajendra Prasad,** Professor, Centre for Rural Development and Technology, Indian Institute of Technology, Delhi, Hauz Khas, New Delhi.
9. **Dr. S.P. Mishra,** Vice-Chancellor, Dev Sanskriti Vishva Vidhyalay, Shanti Kunj, Haridwar.
10. **G.K. Ghosh,** Deputy Chief Executive Officer KVIC, Mumbai.

1

Policies for Rural Industrialisation

Kamal Narayan Kabra

Industrial growth in post independence India impressive in many respects, cannot be said to have made a noticeable impact on reduction of poverty and reduction of socio-economic inequalities either directly or through a combination of trickle-down and pull-up.

It did not generate reinvestible surpluses for transfer to anti-poverty sectors—both its public and organised private sectors—are guzzlers of resources transferred from the rest of the economy. It did produce major inputs for the industrialisation and marketisation of agriculture (the essence of green revolution) but with doubtful anti-poverty impact, particularly in the context of the terms on which these supplies were made available to reduce employment elasticity of agricultural growth and rising capital intensity and ecological cost. As a result, agriculture and industry still remain inadequately integrated. Growth of industrial investment, output and its diversification have not been able to provide productive participation (to avoid the inappropriate term often used in this context, viz., employment) to affect any transfer of labour force from the primary sector and correct an endemic imbalance in our occupational structure. Thus neither wider, alone equitable, sharing of the fruits of industrial growth nor endogenisation of its basic motor power have come about. Low and declining labour-absorption in green revolution, i.e., agriculture as well as organised private industry and increasing capital, foreign debt, energy and import intensities of efforts at maintaining the tempo of such industrial growth put a serious question mark on continuing with more of the same as far as our development strategy is concerned. These factors point to the need for searching methods of developing the rural non-farm sector in a big way for improving levels of participation and sharing in development. What is maintained is that the place assigned to 'modern' industrialisation and policies followed to attain it have not only created a wide chasm between the social and economic aspects of industrialisation but even its 'economics' and sustainability have become highly problematic. Let us note the main elements of the approach followed to date:

1. Creation of domestic capacity to produce capital and intermediate goods, mainly by means of import-substitution.
2. Increasing role of the public sector in the basic and heavy industries.
3. Large and growing role for a controlled and regulated private sector in modern industries with flexible demarcation of spheres.
4. Liberal and multifaceted help for large and medium industries in the private sector within the framework of the Industrial Development and Regulation Act, with special accent on spatial dispersal in favour of backward regions. In fact, the regional and spatial emphasis was in terms of backward areas and not rural areas.
5. As a parallel and subsidiary element, initiation of a number of regulatory, promotional and preferential measures to support the growth of village and small industries (VSIs) basically to subserve some social objectives like employment, dispersal and decentralisation are essential. It could be done particularly to take advantage of their low capital, energy, infrastructure and import requirements.

The incremental attempts to modify this approach to VSIs initiated since the mid-sixties have now reached a qualitatively different phase in which owing to the enhanced value place on modern industrialisation hardly any other element is likely to survive. Even the special dispensation for the VSIs is under threat. The purpose seems to be to pursue modern industrialisation with frontline and sunrise technology even more vigorously in the name of competitiveness and efficiency.

Obviously, particularly the village industries component of VSIs is facing a certain threat of down grading. The concept of 'tiny' sector is just not capable of capturing the essence of rural industries. Rural industries are essentially a component of the informal sector, consisting of micro entrepreneurs who are dispersed all over and represent decentralisation of industrial activities. It is a shallow and lower form of industrialisation—an extension of proto-industrialisation. A few of these industries may take a plunge into higher form of industrialisation. But the vast majority of working poor cannot. The 'free' enterprises of a few blocks the economic participation and enterprise of many. This is where rural industries step in for a widely participative industrialisation. In this sense, these industries are truly in keeping with the philosophy, ethos and requirements of liberalisation and marketisation which are taken as a means of unshackling peoples' initiative, latent energies and resources.

Why have we flattered and are facing the wrong direction insofar as rural industries are concerned?

Common Characteristics

The concept of industrialisation as enunciated in our plans and policies, could not provide clearly identified niche for and identity to rural industries. Following the suggestions of the Karve Committee, the village industries were combined with small industries as though they share common characteristics like households of cottage organisation, traditional origins, high labour intensity, relatively greater employment and output per unit of capital, dependence on local resources without entailing capital goods and maintenance of close links with agriculture and allied activities, imports, spatial dispersal, etc.

An examination of our industrial policies and programmes insofar as they influenced the rural industries shows the following:

1. Rural and small-scale industries were not clearly distinguished. The concept of rural industries is distinct from modern smalo industries. But it remained, by and large, unspecified. The stress on "Small is beautiful" was gradually turned into "Small is sarvy" as though size or scale is all that mattered. It was hardly realised that all village industries are not household/cottage units and all

household units are not rural industries. Similarly, traditional village industries have to be treated as stages in the process of transition to "improved village industries" and not as micro or miniaturised versions of modern industries.

Problems of Industrialisation

The problems of industrialisation of backward areas was recognised, but it tended to be treated as synonymous or a surrogate for rural industrialisation. Thus establishment of any industrial enterprise in the backward and rural areas was considered good enough without examining its responsiveness and compatibility with the prevalent economic, demographic and structural requirements of the region or sub-region.

The erection of factory chimneys and sounding of factory sirens were taken as heralding of an industrial era. At best attention was given to the scale of operations, irrespective of sources of entrepreneurial origin, inputs, technology, product, product-mix, inter-sectoral linkages, environmental sustainability, etc. Nor was infrastructural development in its two-way relationship with industrial growth given operational meaning. The Industrial Policy statement of 1977's concept of 'tiny' sector shows such scale-fixation to the neglect of qualitative dimensions. The exercise of identifying "no industry district" amounted to virtual non-recognition of rural traditional industries.

2. Setting up of six separate Boards for important components of rural industries at the central level under different ministries did not certainly contribute to evolution of integrated regional/microlevel approach to rural industries in which mutual linkages between various industries play a crucial policy and planning element. The centralised approach to decentralised industries seems to have been lost.
3. Constitutionally rural industries form a part of the domain of the states but the centre tends to assume large responsibilities and a uniform all-India approach. Financially weak states with hardly any regional state level planning

and absence of micro-level planning, with the states vying with each other to attract modern, large and medium industries led to marginalisation of rural industries. The idea of "economic federalism", was presented in the Industrial Policy statement of 1980, but without making rural industries its key component.

4. Hardly any attempts were made to forge, renew and strengthen the linkages between agriculture and industry either in the pre-Green Revolution approach or in the Green Revolution approach.
5. Owing to inadequate concern with the inadequate understanding of the problems of rural industrialisation and absence of microlevel planning based on regional specifics, problems arising from excessively small and non-viable size of rural settlements were not responded to. Thus questions like industries catering to a cluster of villages and located at a growth centre of mandi town, problems of small town industrial location, role of municipal and panchayat industrial enterprises to generate an extensive network of decentralised, commonly-owned and controlled public enterprises could not acquire an operational meaning. The role of traditional artisans and the conversion of artisans, rich farmers and merchants into small industrial entrepreneurs, etc. remained outside the focus of active policy formulation. An integrated approach to human habitats, involving question of rural-urban, particularly small towns, relations would have a direct bearing on rural industries. The role of archaic practices like putting out system, etc., and the negative role of merchants and money-lenders was barely responded to.

Distinction

In sum, it was not recognised that (a) while all rural industries are small, all small industries are not rural; (b) that household industries may exist in urban/metropolitan areas too (urban informal sector); (c) that output, employment, skills, intersectoral linkages, generation of surpluses etc,. in the rural cottage sector are generally different from those seen in the small (modern) industries segment; and (d) that non-traditional, non-household industries may be found in rural areas but they are unlikely to be very common and that at present stage of Indian development, rural industrialisation has to proceed in terms of small-scale rural industries. At the same time they need not follow small commodity production organisational pattern.

Emphasis on Small Enterprises

Since the dominant concern of the development strategy was with modern industrialisation tailored on the Western pattern, the major operational *differentia specifica* of the village and small industries was taken to be their is size and dispersed nature. In the scheme of Indian industrialisation, the role stipulated for them was transitional and subsidiary though it was also recognised that some village industries could survive as a special marginal category. Survival of many traditional rural industries and their upgradation, like that of handlooms into powerlooms, was not conceived in terms of the survival of the artisans. Emphasis in practical terms was on *small enterprise* and not on *small entrepreneur.*

It means the *sui generis* role of the rural industries, particularly their income, employment, surplus generation, technological upgradation directly in the hands of the relatively disadvantaged and integration of various sectors was not given any operational recognition. In terms of allocation of investible resources, infrastructural facilities, long and shortterm finance and organisational administrative support, bulk of the attention was given to large and medium industries. As a result of clubbing of small and village industries together, the former took the maximum advantage of the promotional and protective package designed for them. The small industries are essentially similar to the large and medium ones except for their greater decentralisation and smaller scale of operations and wider entrepreneurial base. Their capacity to uplift the traditional industries, brings about a structural change in the occupational structure in favour of non-farm rural activities. Attempts made at developing symbolic linkages between the traditional artisan of rural industries and the agrarian economy and improving rural infrastructure for bettering the quality of rural life were negligible.

Minor Subsidiary

In retrospect it can be said that the rural non-farm activities, basically rural industries, were neither appropriately understood at the conceptual level nor were made an integral part of our development strategy. The primary emphasis was on output growth. Consequently, employment, equity and other social objectives were considered secondary, to be obtained by means of progress in terms of increased commodity production, adoption of modern technology and industrial pattern in the image of the experience of the Western countries. Hence, there is little wonder that rural industries were treated as a minor subsidiary concern. Given the size of rural non-farm sector and the proportion of population associated with it, the policies and programmes initiated to help the growth of the village industries could at best be regarded as symbolic and pale attempts in keeping with the tendency of Indian plans to thinly spread the resources over too many things.

Real Beginning

It is true that in terms of the absolute amount, the resources allocated to these industries and the growth of output, employment and export by them since 1950 look impressive. But in terms of the proportion of rural population finding viable and adequate levels of living on the basis of these activities and in terms of their capacity to become self-supporting, it is evident we are yet to make a real beginning.

It is often maintained that many of the limitations of our policies are at the implementation level. The rural industries are a clear exception where even in terms of policy design there have been inadequacies and outright neglect. It seems either we emphasised investment, technology and production in certain spheres or we emphasised institutional-organisational changes. The two were rarely integrated in any programme, be it for agriculture, land reforms, industry or social services. As a result, the concept of industrialisation was taken in terms of the share on industry in GDP and an attempt was made to create, by means of a high priority to heavy, capital and intermediate good industries, the essential wherewithal of a Western style industrialisation capable of giving a product-mix which was imitative. Capital and energy-intensity has increased many times since the days of the first industrial revolution. The size of the population clamouring for employment is many times more than that of the early period of industrialisation and is rising rapidly. Many sources of capital accumulation available to the 'first World' are simply out of the reach of the late-industrialising developing countries like India. These facts are not recognised at an operational levels. Had these factors been taken into account, one would have discarded technology transfer from the DCs to the LDCs as a factor in development. An alternative approach to industrialisation would require either gradual upgradation of the inherited technology or modification/adaptation of the technology obtained from the early industrialised countries according to the needs and specifics of the developing countries. Such an approach would have shown that the pursuit of modern industrialisation leads necessarily to a dualistic pattern with a largely isolated enclave and one-way flow of benefits to the modern industries in terms of cheap labour, low priced raw materials but without any positive impact on the vast of army of labour in the rural areas.

Thus, there is a need to think in terms of an appropriate and multi-dimensional concept of rural industries based on non-farm activities which are closely integrated with agriculture. These industries have to make use of local raw materials, upgrade existing technologies and products, providing inputs for industrialisation of agriculture in a gradual and ecologically sustainable manner. In short, our RD establishment, social science research institutions and planning and policy agencies have to devote considerable attention to ruralisation of industries in such a way that the process of upliftment of the non-farm economy weilds together the economic, social and ecological objectives. It would mean changing the proportion of fixed capital to working capital, product-mix, energy and raw material use pattern and improvement in the tools and machines used in the process of production. They should be used in such a manner that industry does not appear as an alien and overpowering instrument which displaces impoverished and weakens the bulk of the rural population and continues to keep a scarce landbased activity like agriculture the predominant means of livelihood and thereby perpetuate the traditional inimical pattern of social and economic existence and power. Had such a scheme of non-farm activities, gradual upgradation and modernisation in an integral manner been designed as an operational concept with associated policy and organisational instruments, the role of industry and its various segments would have changed the pattern of priorities in our strategy of development. As mentioned earlier, the lingering colonial legacy and the dominance of output growth as dominant concerns did not permit the pursuit of such a development strategy.

Ruralising Industries

In order to industrialise our villages, (i.e., to set up non-farm economic activities which make use of locally available natural and human resources and traditional skills with a view to give a larger flow of much needed goods), we have to take note of the demographic, resource-endowment, organisational, financial, market and infrastructural profile and constraints. Any exogenously given answer which ignores these parameters becomes politically, economically and ecologically non- sustainable.

On no count, be it technological competitive efficiency, political clout of the better-off, non-availability of readily usable alternative technologies, and capacity to design and fabricate machines and tools and provide power and other infrastructural facilities in our far-flung, dispersed and excessively tiny villages, can one justify the neglect of the need to industrialise rural India? *But rural India cannot be industrialised without ruralising industries,* i.e., making them responsive and appropriate to the needs and specifics of rural India. There are some obstacles like majority of rural population living in habitats with less than two thousand persons, which have to be taken care of by taking a cluster of villages or a nearby mandi town or any other small town as the locational centre for some of the rural industries.

Narrowing the Distance

Present day income levels of the majority of ruralites fail to give them any worthwhile entry into the market for industrial goods. Without publicly funded and organised, widely dispersed, regular and guaranteed income earning opportunities on productive rural work programmes, rural industries cannot take off beyond marginal levels. These programmes can at the same time be need for providing rural infrastructure facilitating rural industrialisation.

An important aspect of rural industrialisation is that while it diversifies the rural economy, it narrows down the distance between agriculture and industry. *Essentially, rural industrialisation involves industrialisation of agriculture.* Green Revolution could not go far in this pursuit because the inputs like fertilisers, pesticides, irrigation equipment and energy came from urban modern industries with practically no involvement of farmers, ruralities and their local resources. The backward linkages of Green Revolution essentially skipped rural India. Even for HYVs, Corporate entities and MNCs took the major responsibility.

Rural industrialisation involving agriculture as an integral element will lead to following developments:

(a) Agriculture is not confined to production of grains, pulses, oilseeds etc. Horticulture, animal husbandry, dairying, fisheries, sericulture, production of organic and bio-fertilisers, non-conventional sources of energy, use of agro and bio-waste, etc. are treated as components of agriculture or composite rural agro-industrial complexes. Provision of drip and sprinkle irrigation, vermi-culture, watershed development, etc. are agriculture related activities which will help industrialise rural India.

(b) Appropriate, post-harvest, value-adding, agro-processing technologies have to be evolved and adopted. Just as modern industry reaps many economies and internalises external economies by integration of downstream and "upstream activities, rural industrialisation based on industrialised agriculture has to adopt mutatis mutandis backward and forward integration. These changes—technological and organisational—would end the present sharp differentiation of agriculture from industry, and make them closely integrated. It has been shown that flexible specialisation can be adopted in organisational, institutional and technological senses by household tiny enterprise based small but inter-related rural industrial units. This opens up immense possibilities for the viability, strength and extended replicability of small industrial rural enterprises.

The time is ripe with liberalisation of the planning and policy regimes to make rural industries an important element of our development strategy and go ahead with the nuts and bolts of these programmes. Rural industrialisation contributes positively to structural adjustment, though the latter may make things difficult for the former being an essentially in ward-looking process. A final point: there is no conflict between people centric rural industrialisation (which changes the structure and nature of the economy insofar as livelihood security for *everyone* and integration of diverse activities one symbiotic basis are the guiding principles) and modern science and technology, of course, with the caveat that one goes in for appropriate, and not blindly imitative technology.

2

Technological Progress, People's Development and Liberalization

Kamal Nayan Kabra

A major basis of the progress of human society has been knowledge, especially functional knowledge, i.e., technical knowledge. In the last five centuries advancement of human skills has taken many great leaps forward, which resulted in transformation of the social sphere leading to industrial revolution. The industrial revolution and the new social set up have resulted in completely changing the face of human civilization during the 19th century. We may now be said to be entering into the next phase of this ongoing social transformation as yet unfinished. This continuous process has led to a belief that technical progress is an inevitable destiny of human society. But just as the life of human society and its transformation is multi-faced and many sided, there are many aspects and many facets of knowledge and technology as well. Generally, advances in the functional use of natural sciences in productive and economic activities are seen as sign of technological progress. But is this process entirely organized on technical or economic or financial basis, or should it even be organized in this way.

Those who always advocate the immense and often miraculous possibilities offered by science and technology base themselves on a sound basis. Most of the achievements of the modern age and their extensive usage are a gift of science and technology. But they cannot guarantee the immense social benefits that could be derived from the use of technology. It is even more difficult to determine for whom, by whom and for whose benefit technology ought to be used. Take the example of railways, telegraph, radio, electricity, atomic energy, telephone—none of these have ever been used by all the public, nor used in a way that would benefit all the public and certainly not in a just rational manner.

In reality, not every new stage of knowledge can be transformed or developed into functional technology. Individual interests, sudden influences, economic necessities and pressures all play a decisive role in transforming inspiration and knowledge into functional technology. Functional technology, the organizations which use them, and their experience, can all have a deep impact on the progress of technology. The international status or balance of power, their internal situation, the reserves of energy and available resources also have a role to play in giving new direction and shape to their technological progress. While on this issue one must say that technological progress is neither fully based on the attainment of available knowledge nor does this knowledge alone decides the course of technological progress. In other words, the boundaries of the autonomy in the area of technology are fairly limited.

Today much of technological knowledge and innovations tend to become the personal or corporate patented property of those who have discovered and patented them. The owners of many of these new innovations happen to be big companies. Just as these companies are engaged in production of commodities and services, they are also involved in research and development (RD), so that they could sell these and earn more profit, sole *usufruct* rights and financial economic gain. In this way, most multinational companies (MNCs) have their own complex socio-economic financial objectives. These objectives are always contrary to the interests of the neglected weaker poorer nations. One can agree unhesitatingly that as vast complex, and multifaceted an issue as technological development should never be completely in the hands of technical specialists, nor private multinational, multipurpose corporations, nor a few chosen international organizations.

The control or decisions about the direction and content of technology and its development should not be left to any one part. There must be wide and democratic debate on the issue. Otherwise, the glaring social inequities and deprivation that we have witnessed over the last 500 years, the neglect of wider social interests and promotion of a few selfish ones, and the callous attitude towards human, cultural and environmental values, will continue to grow. We are compelled to encounter all these in form of industrialization, profit oriented market system, colonial wars, violence, poverty, and total weakening of various nations or you may call it, suffering the consequences of distorted development policies. It is not as though the scientific and technological revolution has been only a bane. Basic comforts and amenities of life have been available to many. Admittedly, however, the available benefits have kept pace with the possible benefits. As a well-known economist has said—there is only one thing worst than the capitalist exploitation—and that is, not to be a victim of such exploitation. In fact, those who encash the profits from the achievements of technological possibilities, play a decisive role, in deciding the social system and maintaining the balance of power.

Lack of co-ordination and systematic social management of the relationship development has led to a widening of the gap between the immense possibilities offered by science and technology and the ground realities in terms of their practical implementation. Technological advances seem to be used to serve limited vested interests. Although it must be admitted that these advances have led to lowering of birthrates, spread of education, improvement in standards of living and forging of better relations between nations and within the countries themselves, one cannot overlook the benefits of science and technology merely on the basis of failure to achieve all the possibilities offered by it. Similarly it would amount to a blind move if we were to move ahead without rectifying the hitherto committed wrongs.

How can one ensure participation in technological progress (as creators, users and beneficiaries) of the common citizen, the villager, especially the poor, landless, small and marginal farmers, small self-employed craftsmen, adivasis, dalits and other deprived sections of people? How can one ensure that they stand to benefit from this revolution as well? Is it possible that the technological advances can be used for development purposes without the direct involvement of these sections of people or without addressing their problems, or researching on them and handing over the results of these researches to them. This fundamental question is a legacy of the last two-three hundred years and has now assumed serious proportions. Certainly, conventional technologies have transformed the lives of the average Indian; but the benefits have only been a small percentage of our investments in it. Would the industrial revolution from the eighteenth century upto now have been possible without the contribution of India and other third world countries in terms of labour existing product and process technologies, capital market, business and natural resources? Whether there be any casual relationship between the resources looted from the third world countries and the industrial revolution or not, whether there be any differences of opinions and the opinion on the moral sanctity of this or not; no one will deny that the tremendous exploitation of these countries played a big role in the spread of this revolution. It is also true that the nature of this revolution and the way it spreads overlooks the need of the colonies. But it may also be said that just as the industrialized nations have burdened and exploited generations of the colonized people without giving them their due benefits, the young, prosperous, and powerful sections of the new generation in these countries have also been playing a similar role. They were not able to take advantage of the industrial growth and help in augmentation of employment opportunities or national income. Sometimes they joined hands with a few foreign powers, or sometimes with ongoing aggressive protests against government policies. Neither of these sections ever succeeded in using technological progress for social development based on people's need and people's participation. They have in fact, always been obstacles to these. Before understanding the role of technological development of Gram swarajya and rural industrialization, one must try and understand the original form, historical experience and characteristics of modern technological progress.

First of all, technological knowledge is the product of the largest social situation and through the passage of time becomes an important tool of social change. The concept of an unbiased (neutral) balanced objective science and technology is an illusion. Those who control the development of technology are generally its primary and major beneficiaries. Technology imported from outside should be adapted to local situations, to enable them to solve their problems. It is unscientific and harmful to blindly believe that modern technology per se superior and is the key to development. Technological innovations through research and development have been rendered a saleable commodity. Technology born out of narrow selfish objectives fosters and results in cutting oneself off from ones local and national problems and leads to maldevelopment and dependence.

Those processes, eg., activities or goods, in the production, and consumption in which the common man has no participation, or which do not provide them employment or regular how of adequate purchasing power, are of no use to him. Those processes which lead to exploitation and gradual exhaustion of natural resources, which spreads pollution, and which gives some people an existence of prosperity and others of deprivation, and which neglects the needs of the present and future generations, can only result in mal development, misery and injustice. Those who paint a rosy one-sided picture of the double edged swords of science and technology, and consider that moulding ourselves on the pattern of the so called developed countries as the ultimate goal, turn a blind eye to the difficulties, problems, emotions, ambitions and needs of the common man. They seek to fill their own coffers.

In the last decades, poor nations of the third world have been heavily burdened by foreign debts and also the effort to gather resources for their treasuries. To help them overcome these crises, the richer nations have pushed them onto the slippery path of the liberalization.

The apparent objective of this externally dictated programme was to enable poor nations to enter the path of scientific progress, so that their products could be sold to the richer nations at competitive prices. Together with industries, basic services, these technologies have been presented as tools to achieve poverty eradication, ecological balance, agricultural development, social services, economic equality, rural development, people's empowerment and quality of life. Advocates of modernism have projected the technological knowledge of multinational companies based on open economy, open market, privatization etc. as the answer to India's problems. Growth and export, increased production, international competition, spiraling cost of economic production (often termed as development costs) poverty eradication, spread of education, food security, ecological balance etc., health improvement programmes will all put underdeveloped nations on the road to fuller development.

Today, when the scientific-technological revolution, is riding on the crest of the information revolution, international organizations like the UNDP are considering new paths to human development on the basis of information technologies and biotechnology. The 2001 Human Development Report indicates how technology is going to be used for improving the quality of human life.

On the one hand, the report supports the use of science and technology as the magical answer to our needs. On the other hand, it advocates pro-people, pro-poor policies. It is not clear which is the main voice. However, the report's propagation of science and technology as the solution to the problems of poverty and as the ultimate "mantra" for development contradicts the report's own statements at various other places.

We can cite a few examples to prove this. The report says that the cost of installing and operating a telephone in India is so much that only two or three per cent can afford it. At present only about one and a half-crore people are able to avail of this facility. If this were to be made available to about 15 to 20 crores of people, i.e., only one-fifth of the population (and still leaving about 80 per cent), then the cost of installing a telephone would have to be brought down by about 50 to 65 per cent. One may say that a telephone is beyond the reach of an average Indian. When this is the case with such an old technology, then one can say that all the latest technological advances in cyber and information technology, motor cars or even less advanced ones like electricity or bulbs or fans will be far beyond their reach. When even obtaining two square meals, healthy, nutritious food, safe drinking water, a roof over ones head etc. still remains the distant dream for many, how can we possibly expect that modern, often "borrowed" techniques, highly sophisticated and complicated technologies, without the participation or empowerment of the people, will be effective tools for their improvement.

This does not mean at all that we accept the situation as it is, or reject modern technology outright. To make the common man, especially one living in the villages an active participant in the technological development is an essential component of a development process that respects the human dignity. But to believe that our technological knowledge is almost nil, and we cannot make any progress in this sphere without the help of large multinational companies and their counterparts amounts basically to foster dependence, inequality and narrow selfish interests, which is not only unjust, but self-destructing. What is needed is to promote one's traditional technologies, especially those rooted in the daily lives of the people, and to recast these according to the present day needs of the people. It is necessary that prejudiced mindsets against public policies and institutions be prized open, and policies for technological progress may be instrumental in achieving this.

The question is directly related to making technological progress a part of our total development process based on policy of collecting self-empowerment. Technological progress, though important and necessary, is in itself not sufficient. In the absence of people oriented, egalitarian policies technological progress only becomes a thorn in the flesh which cannot be easily got rid of. Powerful vested interests also prevent the benefits of technology from reaching every village and every home. Such unsuitable technologies may bring economic advancement but still pose an obstacle to suitable development. As the UNDP Report on Human Development makes it clear: in the absence of people–oriented policies, new initiatives in technology, instead of bringing fellow travellers on the road to development, become tools to deprive them. That is why this report states further that if we are to remove these inequalities in the technological sphere, and hope to attain a position of equality, the poorer nations must be careful in choosing their strategy. They should enter only few selected fields on the basis of their limited tools and resources. But even after all this, one cannot ensure a people-based, technological development. Emphasis on information, communication and Biotechnologies, neglect traditional technologies, declining employment opportunities in the wake of liberalization, increasing inequalities, anti-poor policies, has minimized the adoption of modern technologies, by large section of the people. There is no mention of all this in the report. On the other hand the report seems supportive of the WTO's policies on Intellectual Property Rights (with some amendments). The report therefore contradicts itself in many ways.

Not only India's experience of liberalization, but similar trends in the old policies and programmes have clearly shown that the reliance on technology as such has not been able to ensure even a minimum livelihood and security for the common man, let alone empowerment. They have not been able to give him the tools to refashion traditional technologies according to the needs of the present. Nor have they been able to revive old handicrafts or lay a strong foundation for village industrialization, a major component of the process of industrialization without which India cannot be industrialized. The green revolution has clearly shown us the dual nature of technologies, incompatible with the tools, resources and skills of the majority of the farmers of India. No doubt, we have taken a great leap forward in production of foodgrains; but millions of families continue to wallow in hunger, malnutrition, food insecurity and fall in social stature.

The decade of liberalization has seen neither an increase in rural employment (in fact, the organized sector has seen an average decline) nor in the number and strength of units in the unorganized, informal sector.

The rate of increase in earnings of agricultural labour has been only marginal, which is nothing in comparison to the rise in other sectors. The multinational companies are busy appropriating traditional technologies and other resources as their own. And without increased participation in economic and social life, the question of increased participation in technological spheres does not arise. With multinationals making a marked presence in every area, local and indigenous factories are gradually closing down due to liberal import policies. Our economy, instead of making right use of our resources, is getting embroiled in one scandal after the other.

Government budgets are hit by shortage of funds and resources. There is no augmentation in income and internal markets. Foreign capital is being utilized for the purchase of old shares and investing in old established industries. Savings and investments seem to be at a standstill. There is no real and effective decline in poverty. At the policy level imitative industrialization is increasing in large companies (Indian as well as foreign) are still considered leaders and planners in the development process. Thoughts on people-oriented development have been reduced to a minority voice, the paths to rural development and poverty alleviation through technology; appear to have become more and more difficult to traverse. The hollowness of these claims is becoming increasingly apparent. And this awareness is resulting in stray individual attempts at becoming self-reliant. A faint ray of hope has begun to light up the otherwise the dark horizon, initially at the thought level, but to some extent, at the level of implementation as well.

We must heed Dr. M.S. Swaminathan's timely warning, "it is important to understand that if the market is going to decide the direction of our investment in the research, the orphans will remain orphans and the technological inequalities will only grow wider and wider".

3

Technology and Human Development

Laxmi Dass

Science has a potential to change human sorrows into pleasure and drudgery into joy, if used with an abiding concern for human good. It can create weapons for mass destruction, at the same time; it has the capacity to save human lives by inventing human friendly devices. The result of science depends upon the ways it is used. Science and technology are the most important factors for the overall development of society. It can revolutionize the lifestyle of the people. Today science has made this world a small interconnected unit.

When we feel proud of the outcome of science, we also see a vast area either untouched or half touched by it. Cottage sector is one such area which requires special attention. There seems to prevail a myth about the cottage sector. It is usually considered that science has no role to play in the field of rural industrialization. This is absolutely incorrect. Rather science and Technology for the village industries is as important, if not more, as it is for any other sector. Therefore introduction of appropriate technology miniaturization of technology or upgrading of technology to village industries is an important factor in enhancing their efficiency.

My experience of working with the rural people showed they have tremendous knowledge of their own inherited traditional technology required for their local industry. Small changes here and there with inputs from modern technology can go a long way to solve the problems of rural people.

Let me illustrate with a concrete real life example. There is a tree found in the Shiwalik hills of Himalayas known as Beol in Himachal Pradesh Dhaman in Jammu region and Bhimal in Uttaranchal. Its Botanical name is Gravia Optiva. This is a farmer friendly tree. It is found in plenty in that region. It is a wild growth and hardly needs special effort to be maintained. It has three best F's. Best Fuel, best Fiber and best Fodder. As of today its fodder is used for milch animals but the other two F's are not optimally utilized for the simple reason that to extract fiber from the bark of the tree is a tedious task. If some research is undertaken and easy to operate, maintain and comparatively cheap technique is developed, it can boost the economy of the whole region. Similarly a small diversion to fiber can enhance the marketability of the product. This is just one instance; there are hundreds of areas where a small technological intervention can generate many opportunities.

Ours is a vast country with multi-dimensional problems and resources. We not only have unity in diversity but also diversity in unity. A major chunk of our population is living in the rural areas and is feeding the whole country. Therefore rural reconstruction is the most important factor in order to have balanced economic growth. I don't see any prospects to develop the rural area without having employment for everyone. Industry is one area, which generates employment provided it has a decentralized, people friendly character. Of course, big industries generate employment but only for a few, they take away the employment of many, with uncertain and doubtful net contribution in the form of additional employment opportunities. At some stage the efforts were made to establish big industries in the rural areas. Some industrialists did go to the rural areas. However establishment of big industrial units in the rural area has not injected enthusiasm in the rural people because wherever a big industry is established the villages of the surrounding areas tend to disappear. So far as employment is concerned very few residents of that particular area are benefited. The industry requires skilled labour, which is not available in that area. They are sourced from the other parts and at best some unskilled labour is hired from the local area. Therefore, the actual need is to have rural-friendly industry in which the rural area, which will strengthen but not destroy the village, which will generate employment rather than killing the job opportunities. Now if rural industries are equipped with the latest, generally imported technological devices they can hardly respond to local circumstances. Instead if we tailor-make appropriate technologies which can reduce drudgery, increase earnings and improve the acceptability of the product, the rural masses will be benefited. Therefore, rural industries with appropriate technologies can industrialize the rural area without disturbing the basic rural character.

Another area where technology can play a vital role is to bring industries and agriculture together. The major component of the rural India consists of small and marginal farmers, rural artisans and landless labourers. The landholdings in the rural area are getting reduced day by day with the split in the families. As a result, people in the rural areas are not getting jobs round the year. Therefore, they are in a dilemma. Neither can they depend upon the small time jobs in rural areas nor are they hopeful or skilled enough to get jobs in the urban areas. The situation of small and marginal farmers is rather precarious. They are unable to obtain sufficient income from the small holding they have and at the same time it has become difficult for this category to maintain their agricultural base and also add to their income by working in the urban areas. Therefore, by bringing industry and agriculture together, giving training to the needy and developing industries suitable to the rural people by using latest suitable devices answering their needs one can change the face of rural India. If a small or marginal farmer, landless labourer has four to six months work in his or her hands in agriculture and also gets a supplementary job through cottage industry for the remaining months, this cannot only help the person to maintain his income but also will increase it. To have a technology suitable to develop this system will be a welcome step and can solve the problems of the rural area in a big way. Our efforts should be aimed at developing a technology, which can be introduced in the rural industries sector without destroying its livelihood ecology and values.

But let it be understood that technology doesn't mean any technology. It has to be as per the needs of the entrepreneurs. Technology once developed is not enough. It is a continuous process of RD. This is the basis of the whole process of turning out quality product. However the concern for quality should not destroy the avenues of employment. Quality for whom and by whom? The technology should be introduced in such a way that it must improve the product without negatively affecting other important factors. Our quality control system should fit in with the situation of our country, for this again Science and Technology can play a major role.

Keeping the importance and specific needs and compulsions of rural industrialization in view, Science and Technology has to be given a major thrust. Without R & D, no industry, howsoever small it may be, can survive grow and branch out. The only important thing is that technology should develop in such a way that the people living in the rural area having little facilities can handle the same and make best use of it. Costly, complicated and unwieldy technology will not be accepted by the small entrepreneurs operating in rural areas. Small is beautiful. Let us not in the name of the technological progress, deface and denude this eternal source of beauty and joy.

4

Notes on Intermediate Technology

D.R. Gadgil

(Intermediate technology can be developed by modifying traditional technology, by adapting advanced technology, and most importantly, by experimentation and research to establish it directly. Because of its importance in spreading industrialization, it is worth a major research effort.)

It is necessary to be clear as to what is meant by intermediate technology and what is the place it occupies in the total development plan, especially in relation to the plan of industrialisation. The term "intermediate" denotes a stage somewhere between the "advanced" and the "backward". In India, *the "backward" may be identified with the technology of traditional Indian industry, urban and rural, and the "advanced" with the technology of the industrially developed countries of the West.* The "Intermediate" presumably lies somewhere between the two.

A more concrete concept of the intermediate may be obtained if we attempt to understand why adoption of intermediate technology is considered important in India and how such a technology is expected to evolve or develop.

The need to think in terms of intermediate technology appears to arise primarily from the redundancy of human labour in the country and the inability of the economy to use its labour resources in anything like a full or adequate measure. Important related factors are: the very high proportion of those living in rural areas and engaged in agriculture and other primary occupations; the high rate of increase of population; the generally unfavourable land/man ratio; and the desirability, for better organisation of agricultural production, of reducing numbers currently employed in agriculture. All these circumstances call for the creation, on a very large scale, of *new non-farm employment opportunities*. This can be done on a long-term, stable basis only through the establishment of an industrial base of maximum size and strength. Further, in view of the present distribution of the population and the paucity of capital resources, the industrial complex to be built up has to consist of very large numbers of dispersed establishments capable of creating numerous employment opportunities in relation to capital investment.

A subsidiary, but not unimportant, consideration is the fate of the large numbers engaged, even today, in traditional small-scale industries—rural and urban. The total situation makes it urgent that the route of development be so planned that significant numbers of these will not be rendered unemployed before alternative avenues of employment are opened for them and that, as far as possible, the skills and capital equipment now invested in traditional industry will be utilised in plans for development of industry.

Recent experience gives further support to the analysis of requirements set out above. In the first instance, it has been increasingly emphasized that, in underdeveloped economies, the transmission of economic forces is very slow and partial. As a result, the creation of a small number of centres of advanced industry in a country fails to have any significant impact on the bulk of the rural masses and does not initiate a process of transformation which embraces, within any reasonable time-span, the whole of the economy. In addition, it appears that implanting a relatively small element of highly advanced modern business, including industry, often leads to the development of a "dual economy". This implies the semi-permanent existence of two parallel sectors, one characterised by richness of resources, advanced techniques, and economic progress; the other by poverty, backwardness, and stagnation. Exchange relations between these two spheres would then take on the exploitative character which marked the is exchange between industrialised metropolitan countries and their predominantly agricultural dependencies in the colonial era. It may thus happen that the two diverge and become antagonised instead of merging and progressing together.

Everything thus points to the desirability, nay urgency, of initiating widespread industrial development in all regions of the country to prevent accentuation of dualistic features within the economy and make for concerted and uniform economic progress. It is necessary to set this out because the scientists and the technicians must be made fully aware of what is expected out of the adoption of intermediate technology". Their efforts must be directed towards the selection and development of those procedures and techniques which can serve the given aims.

Its Characteristics

Does the definition of aims and purposes make possible any generalization as to characteristics of intermediate technology? The broadest generalization one can make is about scale. The smallest scale compatible with efficient production is an obvious desideratum, and a significant step in this direction is expected of all intermediate technology. Negatively, intermediate technology would also be expected to keep investment in fixed capital as low as possible; this would be balanced there being no special emphasis on economy in inputs of labour, especially of those types which are not scarce in underdeveloped economies. These conditions arise out of relative paucity of capital, the tendency for fixed capital cost per unit of product to rise with diminution of scale, and the redundancy of labour and its low cost. It is in this context of use of power may also become important.

Because of undeveloped communications and relatively high transport costs, most situations in which intermediate technology is adopted are characterised by cut-up and limited markets. Production units within these would be protected to some extent from outside competition by the transport costs and by any preferences for local products that they are able to establish. The permissible margin in costs per unit of product between advanced and intermediate technology would thus be indicated chiefly by the saving on transport and marketing costs, by any diminution in quality (finishing, packing, etc.) tolerated by the customer in favour of the local product, and by lower labour costs.

An advanced technology is by definition one which makes a more efficient use of resources than a backward one. Therefore, economic development is, in a identified with adoption of more advanced technology sense However, as indicated above, it may be neither possible nor desirable to adopt the most advanced technology available in many areas of the economy; an effort has to be made to discover the technology appropriate to a particular field in the context of both the general and the specific economic situation. Where this process of adaptation has to be very widespread and appears likely to persist for a long period of time, it is possible to talk in terms of evolution or development of intermediate technology or research and experimentation in its behalf.

Discovering Intermediate Technologies

From the point of view of a non-technician, the following approaches to the development of intermediate technology appear possible.

1. One approach may be to start with existing techniques in traditional industry and to utilise knowledge of advanced techniques to transform them suitably. *Transformation* implies retaining some elements in existing equipment, skills, and procedures. If there existed possibilities of a continuous process of transformation so that, starting from the traditional technique, one could, through continuous change, reach the most advanced technology, the problem would be relatively easy. It would reduce itself to that of choosing, for any industry, for a time and place, a particular position on the continuum as the most suitable. In fact, however, such a situation exists in almost no field and each important advance in technology renders previous plant and equipment, for the most part, obsolescent. There may, however, exist limited possibilities of improving traditional technologies so as to keep their operations at least partly viable. This process of improvement of traditional technology

is extremely important, particularly for that part of the transition in which a holding operation for preventing added technological unemployment appears necessary. It is doubtful, however, whether such modification of traditional techniques would be accepted as resulting in intermediate technology.

2. Another approach would be to start from the end of the most advanced technology and to adapt and adjust so as to meet the requirements, of the intermediate. The chief requirements of this, as defined above, are (a) small scale and (b) change in the proportions of capital and labour inputs. The process appears to have two prominent components. First is the analysis of the different units and processes in the total plant, equipment, and operations. The analysis would initially aim at determining the extent to which the task of adaptation and adjustment of technique could be undertaken relatively independently for separate parts of units. Next, on the basis of a given structure of wage rates and availability of various types of labour, the relative advantages of full or partial adoption of advanced technology for the various separate parts could be calculated. It is this analysis that would uncover possibilities of maximising inputs of low-cost labour and of saving on capital. The second component of the process would be the adjustment the technique and equipment, to the extent it is to be adopted, to a smaller or the smallest scale compatible with efficient and economic production. In some cases, the process would also involve adjustment to special local circumstances such as type of fuel or power available.

3. A third possible approach may be to conduct *experimentation and research* in a direct effort to establish intermediate technology. However, for this to be fruitfully undertaken, it would be necessary to define, for the scientist and the technician, the limiting economic circumstances. These are chiefly the scale of operations aimed at, and the relative costs of capital and labour and the scale of their inputs– possible or desirable. Such direct effort at establishing intermediate technology would undoubtedly be conducted

against the background of knowledge of advanced technology in the field. However, it could cover a much wider range of possibilities than the effort through the adjustment and adaptation approach. It should be clear that the establishment of what is called intermediate technology is not a once-for-all process. It is related, on the one hand, to current advance' in technology, and, on the other, to changing economic circumstances, particularly the widening of markets and change in the relative costs and availabilities of labour and capital. Therefore, the evolution and adoption of any type of scale of intermediate technology has to be considered as essentially a transitional stage, though in the existing situation in India this transitional stage, promises to be fairly long drawn out. The continuous changes in the situation to be visualised in the future are mainly, (a) the adoption in some areas of fully advanced technology and (b) the readjustment in intermediate technology itself rendered necessary by the total change in the situation, technological and economic.

Research and Planning

It follows that intermediate technology cannot be considered by itself; it must be constantly and meaningfully related to the total situation. It must be related to both traditional and advanced technology and to the general economic situation, the plans of development, and to special economic conditions within the particular field of application. If one takes the view that today the most important and urgent technological problem confronting India is that of the adaptation of Indian economic circumstances of advanced technology, it is obvious that the bulk of talent and finance in the technological field must be directed towards this effort. The setting up or getting together of a small number of inadequately staffed and financed units for this purpose is altogether insufficient. The main focus of applied research in national laboratories, technical institutes, and large university departments must be on this work. The progress of advanced technology in every field is being adequately pursued in the developed countries; the special adaptations and adjustments required in India are not given, and are not likely to be given, attention in any other country. They must, therefore, obtain the highest priority in our plans. Intermediate technology should become a national concern and not, as at present, be a neglected field assigned to a small number of specialists, set apart.

Intermediate technology is not only closely connected with all technological work in the country but also, and perhaps even more closely, with all planning effort. No plan of industrialisation can be meaningful which does not pay the fullest attention to it. In fact, as pointed out above, the changing circumstances against whose background intermediate technology has to be evolved are shaped by the plans. On the other hand, realistic planning regarding the route and pace of industrialization must be powerfully influenced by the possibilities and achievements in the direction of intermediate technology. There will be a group of activities, e.g., certain basic industries, capital goods industries, public utilities, etc., in relation to which the adoption of intermediate technology may not be considered. For the rest, the effort must be organised in relation to particular aspects and phases of the plan.

I would suggest the evolution of a concept of a "common utilisation plan." By this, I mean taking stock of the total supply of industrial primary and other materials within the country and planning their allocation to various processes of transformation and industrial use. The planning will include both the processing of primary products and their later utilization. In relation to intermediate technology, the planning of the total utilisation of primery products of agriculture, animal husbandry, forest, etc., whose supplies are available dispersed in rural areas, is specially important. The introduction of the concept of a common utilization plan is specially important in the case of such commodities as, for example, bamboo and agave fibre, bagasse and molasses, hides and bones.

The evaluation of intermediate technology and a proper definition of its position and function would also be facilitated by planning in relation to a whole group of activities, such as building and construction trades. The extent to which the utilization of local materials and local labour can be maximized while maintaining efficient and economic production through selective adaptation of technological advance must be established in detail in typical local situations. However, certain broad common approaches could also be evolved if intensive attention is given by teams of technicians to this widespread and universally important field of economic activity.

Evolving and adopting intermediate technology is a dynamic process. It should claim the attention, in an important way, of the ablest scientists and technicians in the country. The effort today is both highly inadequate and narrow. The input of scientific and technical talent has been meagre and sectionalised. While, therefore, the move to call together technicians engaged in related effort is welcome, this will bear fruit only if it is the first step in a much more comprehensive and large-scale plan of action.

Ecerpted from "Notes on Rural Industrialization" *Arth Vijnana*, Poona, India, Gokhale Institute of Politics and Economics, Vol, VI, No.1, March 1964, pp. 9-15.

5

India's Rural Poverty and Possible Solutions

Y.S. Rajan

Introduction

It is rather shameful that after more than fifty years of India's independence, one has to be talking about India's rural poverty which is of an unprecedented scale. Several studies indicate that even urban poverty is predominantly due to the fact of impoverishment of rural peasantry which forces them to move out of villages to seek some subsistence living in the towns and cities. In this process, they even lose the open space or habitat they had in villages albeit without food and other basic amenities. When they come to the cities, they get access to some food though other sanitary facilities including clean water supply still elude them. And they have to stay in the habitats which place them under sub-human conditions.

There have been a number of schemes and programmes which have been attempted by the State and Central Governments and a few foreign agencies to address the question of Indian poverty in rural areas and urban areas. Many of them are independent of each other and have differing criteria. However, the effect of many of these have only been marginal[1]. There are also assessments that the bulk of the government's funding does not reach the rural areas directly due to various leakages in the transitional paths. As India aspires to become a developed country, with its own position in the comity of nations, poverty removal would become perhaps the utmost challenge. Though there are a number of programmes even now officially "operational", some of which having very catchy slogans, it would be impossible to sustain economic development in the country (even leaving aside the humanitarian considerations about poor people) if India does not eradicate poverty totally both in urban and rural areas within a decade. It is not enough to state through various statistics that so many percentage of people have been lifted above the poverty line. Author is aware of the fact that there are many intricate economic models and statistical techniques which have been developed or used to measure not only to poverty line and also as to how close people are near the poverty line or above, etc. However, author would like to take a simple view that while all these complex models and resultant numbers have their own requirements and validity, the stark fact before the country is that for decades we have not been able to solve the crucial problems facing many millions of our people: to have a productive work and earnings for people to cover their food, clothing, habitat, health, education of their children and preparing them for a better future. The numbers of such deprived people does not seem to be decreasing[2]. There are only many debates over the poverty line.

In a simple view unless the quantum of money available in the hands of people of the poor families increases very substantially and unless it continues to have some growth even in the subsequent periods, we cannot say that we have tackled the problem of poverty. The author has addressed these aspects to certain extent in his earlier two writings. One is "The promise of new technologies for poverty alleviation"[3] and another "A Vision for Rural Transformation: Science and Technology Based Initiatives"[4]. These two write ups provide a systemic view. Several elements of these suggestions, have been successfully tested in a number of parts of India which will be described briefly in this study.

Therefore, the main thrust argument of the author is that we have to introduce selectively various forms of technologies (knowledge/skills) into the lives of poor people in villages and also provide them with various forms of connectivities in order that they cut the vicious circle of poverty and further impoverishment as the country marches towards market driven transactional economy.

When one talks of technology, it is not a uniform solution of one or two types knowledge and skill requirements (a la IT, BT syndromes) but they are at various levels and sectors and would also require investments along with the same. In the process, the poor will also invest but at levels they can afford but get quick and better returns. This is done because they cannot depend upon the subsidy; already the public exchequer is under severe pressure. Also because in the past it has always been the experience that from what is proposed to be given as subsidy through public system, only very little of it reaches the poor. Therefore, the solution ought to be in terms of providing from public resources or otherwise, the right type of knowledge, skill and technological inputs and certain minimum infrastructure for connectivities and knowledge/skill inputs. These can help them to come up with their own investments or attract others' investment. However, giving connectivity and knowledge/skill inputs is something which ought to be done with funds from public system but executed differently than what has been done during the past 50 years.

In yet another paper by the author which is to appear in a journal "Towards a Knowledge Society in India — Issues for Management" 3 he has addressed this issue as to how different levels of knowledge and skills are existing in India and as to how one has to implement strategies to move these various levels to increasingly higher levels, that is, from the existing levels to better levels in terms in value addition. Some of the experiences of the Technology Information, Forecasting and Assessment Council (TIFAC) for enhancing the wealth in the hand of rural poor through Vision 2020 Mission Programme on Agriculture which led by the late Prof. S.K. Sinha at Bihar, Deoria (U.P.), Uttaranchal, Kanchipuram and other areas are briefly described as examples. Before describing some of these details, it is important to understand the effect of globalisation which is impact most economies and societies.

Effects of Globalisation

Globalisation of the economy and the resultant liberalisation within the economic system of the country cannot any longer be reversed. These forces have emerged due to the complex nature of the interaction between the growth of economies of most countries, the resultant trade and market patterns which are unfolding. Both of these (i.e., economic growth and trade and market patterns) which, to a great extent, have been shaped by the nature of scientific and technological changes. Increasingly in all the products and services, the scientific and technological intensity will continue to increase and the rate of change will be such that the elements of scientific and technological contents will continue to change. This means more and more investments in technological innovation and the market processes. Which, in turn, means that within a short period, business organisations have to reap the benefits of their investments as otherwise the technological changes introduced by the competitors would displace them out of market and therefore business. Therefore, their aim is: More and more investments in technology and innovation in shorter and shorter periods and reaping the benefits within the shorter product cycles. This process necessitates a much larger market size than what business organisations used to be content within earlier years. In other words, every firm in the world has to think in terms of reaching a global market even though it may be producing only a part of a product or a service. This is what is called an integration of the production process.

Figure 1 in this chapter describes how globalisation and market forces are linked to rural sector in India. The right hand side of the figure shows the global market which transacts business through export and import. It involves most of the people of the developed countries, and most of firms and business houses of developed countries, as well as a part of people from many other developing countries and firms and business entities which deal with export and imports form these developing countries.

The green circle with right hand end of Fig. 1 (in the side there are few dots representing rich Indians) represents the global market. A few rich Indians have already fully integrated

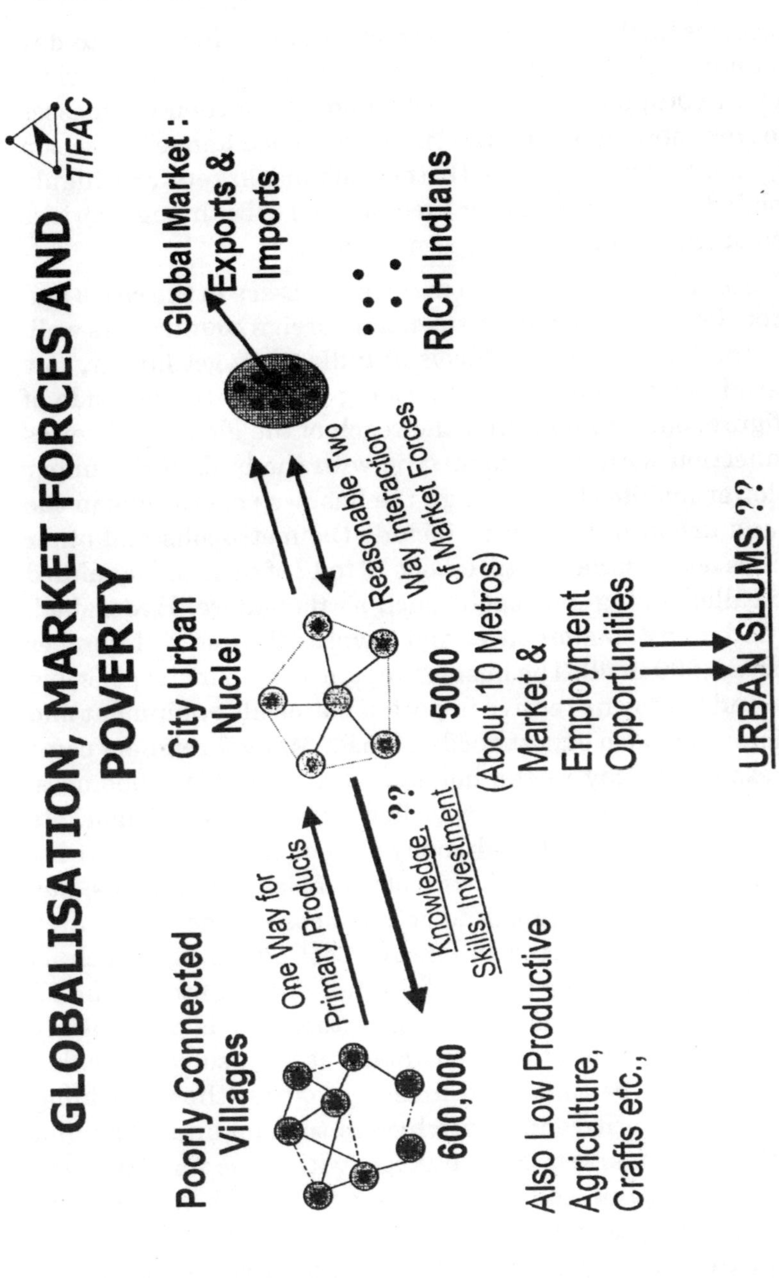

Figure.1

Visit us at www.tifac.org.in

themselves with the global market because their day-to-day consumption material including cloth, food and even drinks let alone the equipment/machinery/technology for their businesses come from abroad. Given the "free foreign exchange" situation now, their transactions with the global markets are highly simplified, (even in the earlier times, it used to be through various forms of legal and not so legal methods.)

The rich Indians control many businesses in the cities and metropolis. Some have investments in foreign countries as well. From the lower economic levels of India, they get labour, raw material, space to operate, etc. Going towards the left side of the figure, one can note that the reach of the global market as a connection within the country is with the bulk of the upper and lower middle class through the business operations in the city and urban nuclei which include the metropolis and other big cities not so declared still (about 10 to 12 of them still qualified to be called as big metropolis such as Bangalore, Hyderabad, etc.) and many smaller cities and towns. All these will number to above 5,000 in all. The business operations in metropolis, big cities and big towns receive imported materials equipment and services; they also export goods abroad. Many foreigners come to these cities stay in the hotels and also transact business. There is a reasonable two way flow between global markets and these city nuclei, though many of them feel the competitive pressure — loss of jobs, closing of inefficient businesses, etc. Other small towns compete to do business and get integrated with these market forces by offering cost advantages such as low cost labour, lesser rentals for office spaces, etc. In the IT fields and for some manufacturing sectors, this is particularly relevant. These small towns etc. make at least some low value added items to be consumed in Indian cities (thus providing them cost competitive inputs) which in turn maintain the link with the global markets. Such linkages with these smaller towns (with the low value addition to products and services) help to maintain a large workforce in the cities and the metropolis thus helping them to continue with their competitive advantages in the global market through cheap labour content and therefore lower costs. Thus people in smaller towns and rural areas (we will expand on this later) still "subsidise" Indian cities with cheap inputs. A part of this "cheapness" of the Indian labour and Indian cities and the resultant cost advantages to Indian businesses in big cities comes by virtue of the fact the urban slums absorb a large amount of human misery about which we have discussed before. They supply the very cheap labour force in terms of house servants, hawkers, cleaners, casual workers, etc. who in turn maintain the "competitive advantages" of the workforce of bigger businesses and the firms to face onslaught of global markets. We need to recognise the fact that since the technological sophistication of Indian industry in general (though there may be a few exceptions) is still poor, much of the competitive advantages in the global market are coming from very low wages of Indian people. How long such a situation can be sustained is a serious question. The author has addressed these in his book[5] and also in the book[6] "India 2020: Vision for the Millennium" by Dr. A.P.J. Abdul Kalam with the author. In the Fig. 1 under discussion, the global markets and the urban integration with the global market are now largely sustained by urban poor, smaller towns and the misery of the rural poor.

Now comes the question of what happens to the poorly connected villages shown in the Fig. 1. At the left end, which are about 6,00,000 in number. Mostly people in these poorly connected villages live on a low productivity agriculture, cattle rearing, and related casual work. These are more or less primary products which find their one way to smaller towns and cities going as more or less as a very low value added commodity items. Paddy is collected and milled in towns; in cities higher value addition takes place. Honey is collected in villages and moved to bigger town sand cities where it is bottled and marketed. Low artisanal goods purchased and marketed in towns/cities as value added products. Same with milk. Thus, most of the villages are always at subsistence levels. Poor people in villages do not have the capacity to hold their products to get better returns. They often do distress sale. Of course there are a few rural rich farmers who can afford to hold the produce; they have in many ways integrated themselves with towns and cities. Many such rich farmers have now started to live in towns or are having at least second homes. Some of these rural rich farmers are themselves under economic pressures which are exerted on agriculture by the global economy. They are trying to get out of agriculture and get into other businesses as is happening in Punjab, Andhra Pradesh, etc. But bulk of the rural poor, and the marginal farmers cannot afford this luxury. They continue to struggle in the villages and get whatever seeds or fertilizers, they can get access to within the constraints, and produce whatever they can and sell the produce on whatever prices are offered to them by the markets of the small towns or the middlemen. The same applies to artisans who produce crafts and other products from villages. The situation of landless labour in villages is still worse. Many of them tend to migrate to cities as unskilled workforce. Occasionally a few tourists may go to some of the villages which have some tourists' attraction or have some other forms of archaeological and historical importance. Again at these places tourist industry develops. But they mostly thrive on products produced in cities: potato chips, beverages, even food products and bottled water come from cities. Occasionally some tender coconut may be bought from these villagers! Of course a few get jobs of selling city (made) products. Some small shops may come up which sell these products and may have a clientele from low income Indian tourists. But by and large this trickle down effects to villages are very minimum.

To recapitulate Fig. 1 the global markets are exerting the pressure for economic efficiency on businesses and operations in the city and urban nuclei. They can face these pressures only with high intake of science and technology; either they have to generate them themselves or allow a foreign investor to be a partner or buy from foreign technology suppliers. In this way (i.e., by technological inputs thus obtained) they can maintain their position in economic competition or even improve themselves. A few 100 million Indians perhaps around 300 million Indians can somehow find their way to survive or even flourish against these competitive pressures or even take advantage of global opportunities by learning to live through this competition. On the longer term, though, they themselves may be endangered if Indian S&T systems and Indian firms do not become technologically sophisticated, with at least some of their own internal S&T strengths and organisational strengths. A perennial dependence on foreign technology through purchase or through investment partners would erode their business advantages. In such a case (of perennial dependence) only a few rich Indians (may be about 50 million) may be able to survive global competition because of the strength of their financial capital and trade linkages. And to a certain extent in the educational system which will tend to produce young boys and girls from India to serve foreign markets and some of these surviving businesses, there could be some vitality. But most persons in the city who are now able to survive global competition, may not be able to sustain without S&T and other organisational strengths. As far as the villages are concerned they are poorly connected even now. They are one way source of cheap primary products, crafts etc. on which further value addition is done in towns and cities to get some export earnings to survive global competition (Note that agricultural products and crafts still are good net foreign exchange earners). So what will happen to the villages if this process of exertion of global pressure continues for sometime and the drive for economic efficiency through market forces aiming at international standards continue. It may mean complete depression of villages because at some point of time it may even be difficult for villages to produce agricultural products and crafts to meet the international/global standards of phyto sanitary conditions or designs in crafts etc. World is not static. Countries like China, a number of countries from south east Asia, central Asian countries etc. may produce products which will sell in India (already happening). Then there may be a few corporates who may buy their (i.e., poor farmers') lands in the villages and introduce modern farming. Then only a few of them may be employed at economically efficient levels to produce competitive products. Similarly, many of the traditional crafts with "new look" may be produced by investors from cities or foreign countries with more knowledge, skills and investment. Again only a few villagers may be employed. All these employments and payments will be driven purely by market forces. What happens of the residue of the Indians whose number will not be small. If now in villages there are 700 million with poor conditions, with population increase to say 1.4 billion by 2020 and with cities not able to absorb more than 300-400 million for reasons described before, they will be about 1 billion who will suffer; or it may be 700 million. It is still large number.

Thus, it is very clear that the villages and the people there cannot be left to the forces of global market forces and the resultant internal market forces through liberalisation and economic reforms. In fact even for the city people (middle and lower middle class and urban poor) there is a case that they cannot be left to these forces alone and large focused efforts need to be done to empower them and Indian industries with S&T strengths and the necessary organisational and business management skills. But for rural India which already suffers from severe poverty, situation is already serious. Also their expectations through TV and other media is gong to go up and they can legitimately aspire for more. It is a human volcano getting ready. Such a strong language is used to draw attention to the serious situation.

In order not to have such serious outcome, the solution is generate linkages of knowledge and skills and investments into the villages not merely as small doles or as something which are given in the name of appropriate technology keeping them in subsistence level, but in substantial manner with a new approach so that larger value addition even within the limited investment level possible for them is done at villages. Today such linkages are absent or very minimal as shown as red arrow in the figure. Providing such two way linkages cannot happen in a vacuum. It will require initial public investment in terms of economic connectivity and also in terms of knowledge and skills which are described in Ref 2, 3 and 4. If this is done with a strong will and steadily over a period of 10 years then the economic levels of the villagers will grow to a good level so that they are not merely people eking an existence at subsistence levels, (i.e., "drawers of water and hewers of wood") but those people at who can produce products and services of good competitive quality with each having their own niche in a global chain of production but without being at a disadvantaged position of one way extraction currently in vogue as shown in Fig. 1. In other words, they can master the market forces of liberalisation and globalisation with adequate tools with them to participate as equal partners. The win-win situation which is being talked in general terms for others in the global forums should also include these people rapidly. It cannot be win-win for others at the cost of these large number of poor people. To equip these people with the tools briefly described above becomes at essential necessity for India to become as a developed country and in order that it is not crippled in economic terms if the policy approach is "to extract cheap labour/goods when available and to dump them at difficult times". How to do this, i.e., equipping rural areas with necessary tools with minimum necessary public investment are described here with a few examples from the TIFAC.

It is also important to point out that equipping people with these tools and related investment is far more crucial than the political and legal initiatives of the Panchayati Raj for villages. The rates of changes due to globalisation are fast. And time periods between opening and closing of windows of opportunities and small. Therefore actions of equipping people with tools and investments have to be done speedily. Otherwise there may be irreversible damages.

Technology Milieu

Details of technology milieu are discussed in Ref. 2 and Ref. 3 in specific terms. The foremost of the sector which will have immediate effect on poverty alleviation is agriculture, cattle rearing, poultry and such food related items. The key issue is to provide the farmer with productivity enhancing knowledge/skills. Most of these will come under what is called science and technology, but applied locally through systematic understanding of local habit and preferences and also choosing the produce, which can be positioned in a proper value chain in the nearby markets. If necessary the farmers have to be convinced to change some of their practices through practical demonstration.

The late Prof. S.K. Sinha who led the Technology Vision 2020 exercises of TIFAC on Agriculture and Food (Technology Vision 2020 document on Food and Agriculture) during 1994-96, had later led the projects which precisely aimed at such demonstrations. The successes achieved by him and his network of scientific teams and villagers are very encouraging. These are briefly described in the following Technology Vision 2020 Mission Projects on Agriculture:

Technology Vision 2020 Document on Food and Agriculture, brought out by TIFAC during 1996 visualizes that the following items need to be addressed for the agricultural prosperity of the country in next 20 years.

1. To attain food and nutritional security of India by the year 2020.
2. India to aim to be a major player in the world in the agriculture sector and a leading exporter of grains and other agri-products.
3. Eastern India to become a major producer of wheat.
4. To broaden the production area in Eastern U.P., Eastern M.P., Bihar, Orissa, West Bengal and North East by introducing modern technologies so as to meet the demand of foodgrains and other commodities by 2020.
5. Rice producing areas to use hybrid seeds on a large scale.
6. Central India to be made a centre of vegetables, fruits, pulses and coarse grains.
7. More emphasis on tuber crops.
8. Water as a national resource — water management as the key to agriculture prosperity.
9. Core post-harvest technologies to be mastered and disseminated.
10. Steps to educate farmers about what is happening elsewhere, if needed provide them the opportunity to travel, and use of space technologies to facilitate interaction and encourage farmers to ask questions and share experience.

To demonstrate achieving of the objectives set in the document TIFAC started Vision 2020 Mission Projects on Agriculture during 1998. Till now TIFAC has launched seven projects in the states of Bihar (Sone command distributaries), Orissa (Nayagarh district), Eastern U.P. (Deoria), Tamil Nadu

(Kanchipuram), Chennai (NAF), Uttaranchal (Pinder Valley) and Sikkim. A few success stories already achieved so far are very encouraging and a brief report about them is given in the following:

Some Success Stories

Approach Followed in TIFAC Projects

- Identification of farmer's constraint on the agriculture as a whole and productivity upliftment in particular by thorough interaction with farmers, field visits, etc.
- Understanding of the existing management practices of different crops
- Proper documentation of the base line data
- Determination of soil fertility status
- Action plan decided in consultation with the farmers without making drastic changes in their existing practices
- Training is an important component. Extensive training cumdemonstration are being given to the farmers before cultivation of each crop
- Proper follow up of each step of management practices
- Few farmers were brought to Delhi, Punjab, Haryana to visit reputed agriculture institutes and farmers field so that can get an idea how others progressive farmers are doing agriculture
- Arrangement of good quality right variety seed for the farmers from Delhi and other areas on cost to cost basis; the transportation cost is being borne from the project.

Bihar Projects

Location: Sone Command Distributaries of Bihar, which is about 40-45 km towards south-west of Patna.

Success Achieved So Far

- The project started in the year 1998 with base line survey of R.P. Channel 5 and Malhouli distributary of Sone command area.

Implementation started in the year 1999 in R.P. Channel 5 and Malhouli distributary.

1999

Area Coverage

Demonstration of paddy in 2.4 acres land, farmers adopted this technology with their own inputs in about 32 hectares. In wheat crop the area coverage was 16 ha, yield was more than double.

2000

Area Coverage

- 432 ha in paddy crop and 1000 ha in wheat crop.

Yield

- Yield was more than double both in paddy and wheat crop

2001

Area Coverage

- 1200 ha in paddy crop and 1300 ha in wheat crop.

Yield

Same trend like previous years, i.e., more than double the traditional yield.

Other Activities

- Zero till machine for sowing of wheat crop has been introduced in this area which helped the farmers to complete the sowing in time
- An effort has taken to improve the dairy farming of this region. A milk co-operative has been started in the village Amwan of R. P. Channel 5 distributary which is supplying about 250 litre of milk per day to Patna Dairy Project. Four youths got trained from NDRI, Karnal on the improved dairy management
- One seed production cooperative in the name of Pataliputra Seed Production Cooperative has been formed. This will be functional in all the distributaries involved in the project

- Farmers have been encouraged to grow fruits and vegetabl. Farmers from different distributaries have been given training on seed production, vegetable nursery raising, maintenance of Agro-machinery etc.

Unique Initiative

TIFAC has planned to take a unique step to replicate the technology in other areas. It has been planned that a few Bihar farmers who had benefited from the project and are ready to go to other areas will be taken to other areas to convince their fellow farmers, to explain their experiences about the technology. This will help in replicating the technology to other areas very fast as farmers to farmers communication is much faster than scientist to farmers communication.

Key Factors for such an Enormous Multiplier Effect

1. Winning the confidence of the farmers.
2. No inputs were given from the project except in first year for few acre demonstration of systems approach technology for cultivation of paddy and wheat crop.
3. The technology is economically viable; farmers are saving money in terms of input at one hand and are getting more output on the other hand.

Latest Initiative

Another important development was that in addition to rapid spreading of interest within the actual farmers, the whole community (in the benefited areas) got involved. Certain joint cooperative efforts for better quality of life have emerged. For example, women group have approached the TIFAC team to help in several aspects other than mere agriculture. Therefore, the project is now being shaped into Integrated Rural Development task. Since women are also get involved in this project, a women Self Help Group is being formed. Local artisanal products like preparation of handicraft items from Munj grass (locally available), silk items are being encouraged. Initiatives have been taken by TIFAC to network this group with an women run NGO "ADITHI" who will help them in providing technology for value addition and also for marketing. Child education mainly computer literacy has been given a thrust.

Benefits in a nut shell

- The yield of both paddy and wheat crop has been increased by 2-2.5 times
- The technology has been multiplied in to about 1300 ha with a span of two and half years
- More than 1,000 farmers are involved in the project
- Farmers are saving money @ Rs. 1000/- per ha in terms of inputs (less application of fertilizer, irrigation, seed, etc.)
- Training to unemployed youth for raising nursery of fruits and vegetables, maintenance of agricultural implements, etc.
- Diversification of cropping systems

Technology as Such

The technologies, which are being followed in these projects, are farmers, friendly, inexpensive, basically optimisation of management practices based on known scientific principles, No modern technologies like Biotechnology, Genetically engineered crop are being used now as these will not suit them now. In a nut shell the technology is a combination of the following efforts:

- Selection of right kind of variety of a particular crop synchronising with the climate, market potential, soil test value, etc.
- Advancement of nursery raising of paddy synchronising with monsoon arrival which has helped in advancement of sowing of wheat.
- Reduction of seed rate in nursery from 30 kg per acre to 10 kg per acre in order to get good healthy seedlings.
- Addition of balanced doses of fertiliser, introduction of micro nutrient specially zinc in the fertiliser schedule.
- Proper irrigation at right time(s).
- Taking plant protection measures.

Eastern U.P. Project

Location: Deoria district which is about 52 km away from Gorakhpur.

Progress Made So Far

- This project follows the same model like Bihar project.
- The project started in the year 2000 with paddy crop demonstration in 27 acre area; within a span of one and half years the technology has spread into 3,000 acres
- The productivity of both paddy and wheat has increased by more than one and half times.
- Farmers informed that the sugarcane production is getting declined since last 10 years. After detailed discussion it emerged that sugarcane seed has not been changed since last 10 years and that is one of the main causes of declining sugarcane productivity. An effort has been made to work on this issue, about 53 quintals of seed were arranged from Indian Sugarcane Research Institute, Lucknow, the entire yield has been kept to use as seed material, which will cover at least 500 acre area this year.
- Farmers are being encouraged to grow fruits and vegetables; about 500 good lichi plants have been distributed amongst the farmers.

Uttaranchal Project

Location: Chamoli District of Pinder Valley.

Salient Features

This project is different. Here, the focus is popularisation of greenhouse cultivation technology, mainly for raising nursery of off season vegetables, aromatic and medicinal plant, as this area is hilly and cold and therefore, crop patterns are different. The approach of TIFAC is to customise the tasks to local areas and markets using scientific and technological principles.

The project has started in February 2001. The work plan envisaged the following:

- Greenhouse establishment for raising nursery of fruits and vegetables
- Selection of site in Pinder Valley for hydro-power (2×10 KW each) and its demonstration
- Drying of apricot for value addition
- Overall agricultural development of Pinder Valley region.

Progress Made So Far

- About 130 greenhouses have been constructed so far covering total area of about 3000 sqm
- Medicinal plants nursery has been raised in 6 greenhouses. The medicinal plants are Cutki, Atish and Meetha
- In other greenhoses, nursery raising as well as cultivation of different vegetables are in progress
- Drying plants of fruits and vegetables has been installed in the village Narayanbagarh and farmers are using it happily. Farmers are using this system even for drying of meat which otherwise they do it in their traditional chullahs. Their traditional system takes lot of time and the meat gets coated with carbon

Training has been given to 11 farmers on construction of low cost greenhouse. Now farmers are constructing greenhouses on their own.

Orissa Project

This project is unique in itself. The Nayagarh district receives a good amount of rainfall, around 1,400 mm per annum spanning over July to September only. Traditionally farmers take only one crop in a year, i.e., paddy. Its productivity is very low, around 1.5 t/ha. TIFAC has taken an initiative to address the issues of rainfed inland agriculture. Rain water harvesting and seepage collection through micro irrigation systems and its efficient utilisation for intensification and diversification of agriculture is the main thrust of the project.

Location: Kommand Panchayat of Nayagarh district of Orissa, 120 km away from Bhubneshwar.

Success Achieved

Phase I

- The project started in May 1998 with socio-economic survey
- Soil testing to determine the fertility level
- Six micro water resources have been created in three villages

Phase II

- Paddy crop cultivation has been demonstrated in 25 acres of land, more than 2.5 times yield obtained
- Farmers are now able to take more than one crop.

First time cabbage is being grown in this plot, normally farmers keep the land fallow after paddy due to lack of water.

Kanchipuram Project

Location: Kanchipuram district of Tamil Nadu

Salient Features

- The productivity of paddy crop in this region is very low (less than 2 tonnes/ha) although this area has potential to increase the yield further.
- Demonstration of cultivation of summer paddy crop and other crops like vegetables, groundnut etc. following the total systems approach technology are being carried out in approximately 30 acres of farmer's land of 10 villages.
- Farmers have been encouraged to grow fruits; a few mango saplings of trees have been distributed among the farmers.
- Improvement of dairy farming of these areas has been given special attention in collaboration with BAIF, Pune. Training of youth for establishment of cattle development centre has also been undertaken under this project.
- New vegetables like snake gourd, bitter gourd, annual muringia etc. have been introduced.
- Farmer are getting better price for the introduction of good quality paddy variety. BAIF, Pune has been involved to improve the livestock status, artificial insemination is one of the activities has been identified.

- Seed production has been give a special thrust.

Sikkim Project

Location: West and South West Sikkim.

Salient Features

- The project started in May 2001 with the base line survey to assess the natural resources, to understand the perception of the farmers, productivity of different crops and their scope for further increase.
- From the survey and through interaction with the farmers it has been emerged out that the target areas are
  - Large cardamom productivity enhancement, which at present is getting declined due to viral attack
  - Ginger productivity enhancement, which at present is getting declined due to viral attack
  - Mandarin orange productivity enhancement

Partness in a Network

Similar efforts are being done under Vision 2020 Projects on Agroprocessing led by Smt. Lila Poonawalla in the milk sector and for cereal/horticulture value addition. A few major projects are being undertaken in the herbal sector to provide local value addition at the farm level so that the farmers get considerable earnings (and substantial returns on their investment) per hectare. It is a question of stabilising the agrotechnologies for the well chosen (market share wise) herbs and placing them in the correct places of value chain. These projects in the herbal sector are sequel to a major report by the Scientific Advisory Committee to the Cabinet (SAC-C) and TIFAC[5].) A beginning has been made in the fisheries sector under the leadership of Dr. S.A.H. Abidi. TIFAC also works closely with BAIF, National Agro Foundation (NAF) and other similar organisations. TIFAC studies in 1993 on leather[6]) have led a major mission implemented by Central Leather Research Institute (CLRI) Chennai as leather, which again helps the poor to upgrade themselves. There other projects in under the Home Grown Technology (HGT) Scheme dealing with use of waste materials as value added marketable products ranging from wood substitutes and natural dyes.[7] (TIFAC websites: www.tifac.orq.in; www.indianpatents.org; www.missionreach.org.in). There is a major mission for technology for bamboo products recently approved. This will greatly facilitate rural poor to earn through selling bamboo with value addition.

These are briefly mentioned because in addition to increasing agricultural productivity at the marginal farmer levels, there are several other recently proven technological initiatives by TIFAC with other partners and industries which demonstrate clearly that empowering rural poor with necessary tools are possible and are successful in further multiplication.

There are many other examples from the work of BAIF[8].

Conclusions

There is an urgent need to empower the rural poor with technological (knowledge/skills) tools for various sectors at different appropriate levels and also crucial connectivities described earlier. There are a number of successful models implemented in the recent years by TIFAC, BAIF and a few other agencies. Networking these efforts is important TIFAC is entering into a memorandum of understanding with Akhil Bhartiya Gramodyog Mahasangh (ABGM) to initiate such a process. It is important to remember that time is of essence. Value addition at local levels (not necessarily total production) is crucial because that alone ensures addition earnings at the villages. The approach should absorb the realities of the market forces of globalisation and liberalisation, so that they can be mastered by our people. Experiments show they can be; but large scale and speedy actions are called for, on the lines suggested in this chapter.

References

1. Kamal Taori, "Understanding Rural Industrialisation, in a book *Sustaining Rural Industries* edited by Laxmi Das and D.K. Giri, Published by Samskriti, New Delhi (2002).
2. Y.S. Rajan, "The promise of new technologies for poverty alleviation", a paper presented in the conference "The role

of new technologies in poverty alleviation and sustainable development" held on 6 November 2000 in Washington D.C., USA, organised by Tata Energy Resources Institute, North America (TERI).

3. Y.S. Rajan, "A Vision for Rural Transformation: Science and Technology Based Initiatives"—Chapter III of the book *Empowering Indians: with economic, business and technology strengths for the twenty-first century,* by Y.S. Rajan, Har Anand Publications, New Delhi (2001).

4. A.P.J. Abdul Kalam with Y.S. Rajan, "India 2020 Vision for a new Millennium", Viking-Penguin, New Delhi (1998).

5. *"Herbal Products — Current Status, Vision and Action Plan",* Report of Scientific Advisory Committee to the Cabinet (SAC-C), Govt. of India (2001).

6. Thyagarajan G., A.V. Srinivasan and A. Amudeswari, *Indian Leather 2010— A technology, industry and trade forecast,* Central Leather Research Institute, Madras, 1994.

7. Technology Information, Forecasting and Assessment Council (TIFAC) websites: www.tifac.org.in; www.indianpatents.org; www.missionreach.org.in

8. "Natural Resource Management for Sustainable Livelihood—Replicable Models", BAIF Development Research Foundation, Pune.

6

Rural Transformation Through Decentralized Technologies: Empowerment and Participatory Systems Management Approach for Sustainable Livelihoods

S. Chatterjee and Sunil K. Agarwal

Introduction: S&T can Make a Difference

India is predominantly a rural country with about three-fourth of its population living in the villages. Sustainable rural development, therefore, assumes prominence in the overall development of the country. Statistics reveals that majority of our rural people are living under sub-human conditions. The problems of poverty, malnutrition, ill health, unemployment, food and environmental security are major concerns in these areas. Three-fourths of our rural population does not have access to proper sanitation, one-third does not have access to clean drinking water, and one-half are illiterate. President Dr. A.P.J. Abdul Kalam has often emphasized that our country has the potential to become a developed country by the year 2020. He has rural transformation in his dream by providing "Rural Prosperity through Connectivity (RUPCON)" through four connectivities viz physical, electronic, economic and spiritual to be brought about in a most creative and cost effective manner. In this endeavour, technology can play an important role, which is well accepted as the main driving force behind all developmental initiatives and an important tool for sustained economic growth.

Technological applications are largely restricted to the urban areas. The rural areas have not reaped enough benefits from them. They must be used judiciously as important tools in the developmental activities to address the problems of rural development in all sectors of the economy such as agriculture, energy, health and sanitation, rural engineering, housing and habitat, etc.

With economic reforms already in progress, our industry and trade are moving in the right direction to reach the developed country status by 2020 provided we simultaneously prepare a powerful rural technology base. We have a very well documented Gandhian philosophy of village industry development. Now, we can employ the most modern tools of information technology, biotechnology and knowledge systems. There is a need to intelligently integrate our ideas and design a management strategy to develop a strong RT base in agricultural and non-agricultural, and infrastructure-cum-service sectors. Setting up of more and more RT based rural enterprises will not only upgrade human skills, generate employment, and create wealth, but it will usher an era of growth, development and other opportunities. This will necessitate a radical change of mindset and a strong belief in innovativeness and management of change concept.

Need

Decentralized Way of Functioning: Good Governance/ Management

In many countries decentralization was the principal institutional development strategy for reaching at the local or grassroots level. In the enactment of decentralization policies, various strategies and structure of local developments were adopted. Yet development was uneven across sub-regions, even within a country. In the high growth economies, opportunities were similarly created by opening the markets at the local level. The traditional cropping and labour demand patterns were replaced by market driven high valued commodities and demand for skilled labourers. This happened in Karnataka, India and central Thailand, where paddy land and mangrove areas were converted to high-valued export earning prawn cultivation. Now, fresh attempts are being made to transform institutional tools and enable people to be participants and beneficiaries of development governance. It is now being acknowledged that institutionalization of development governance at the local level could increase participation of the people in the mobilization and sustainable management and utilization of resources for development in their respective areas. At a number of United Nations forums commitment has been given to the need for empowerment of the poor to eradicate poverty for which good governance is stated to be prerequisite to successful rural transformation. People need to be enabled through education, motivation, empathy and technology support from political and professional communities that work on their behalf.

Appropriate "Green" Technology: Choice as Per Local Needs

In the rural context, it is therefore, necessary to develop and introduce appropriate or so-called "Green technologies" coupled with sound delivery system, which ensures economic and ecological sustainability and optimum use of local resources emphasizing on technology capacity building amongst rural people. In this endeavour, institutional linkages and active participation amongst voluntary agencies/science and technology based field groups, RD institutions, financial agencies and above all people who are primary stakeholders become crucial for improving the quality of life in rural areas to achieve long-term sustainability. In above process, technology choice can have a critical impact on many aspects of rural transformation, especially the way we choose them, the way we innovate and design them and the way we deliver them to masses. Therefore, technology must be carefully chosen to enable rural people to acquire and imbibe knowledge of technologies appropriate to their needs and environment. The technologies should be such as would upgrade their traditional skills and capabilities; minimize fatigue and reduce drudgery; be innovative and capable of easy assimilation; generate significant and assured added value to existing methods of operation; generate employment and use local resources both men and materials; need low capital investment and should result in low cost production of goods; be capable of replication and adoption and should blend harmoniously with existing eco-systems leading to tangible improvements in the living conditions and self-sustained development of the rural people.

Thus, appropriate or green technologies with above features can play a crucial role in building up local capacity, to devise solutions for tackling the identified problems, to improve the lives of the rural people by improving their surroundings and daily activities. Focus must be on technological empowerment of people with skills and critical thinking that fosters a sense of self-reliance and ability to evaluate what is beneficial or detrimental to their interests. This will improve their access to affordable, environmentally sound technologies and generate meaningful employment in local economic structure leading towards poverty reduction.

Rural people can participate in technology in three ways as consumers of technology, as producers of technology and change agents in dissemination of need based technology. Some of the emerging technologies, which are directly beneficial, are energy technology and small-scale biotechnologies for sustainable agriculture, while, others like information and educational technologies offer lots of scope to overcome the barriers of illiteracy, lack of education and participation in development and adoption of such technologies. Livestock enterprise with appropriate production technologies also offer immense potential for social transformation of rural lives. This will help to strengthen both nutrition and livelihood security. In terms of emerging technologies, biotechnology offers a powerful capacity to produce science based products like bio-fertilizers, biocide, microbial culture and vermi-culture for sustainable agriculture; tissue culture based production of fuel, fodder and commercial species of plants; fertility control vaccines for population control; improved production and productivity of cattle and buffaloes using embryo transfer techniques and conservation techniques to tackle narrowing genetic base to meet the needs of rural womenfolk in particular and community as a whole.

Experiences in the past showed that the introduction of technology and mechanization based on top-down perspective have favoured rich with capital assets and displaced millions of landless producers, especially women from agricultural work. Therefore, technology and its scale must be carefully chosen to enable rural people to acquire and imbibe knowledge of technologies appropriate to their needs, priorities and rural environment. An important point is that focus should be on diversification of labour use and not to the displacement of labour. As in the case of agricultural implements for use by women, major factors to be considered are the characteristics of the tools such as handle, and size, based on the anthropometric data on women, the preferred body position, elimination of drudgery and economic enhancement.

Technology Needs and Back-up Support: Appropriate Package for Sustainable Livelihoods

Appropriate Technology Package transfer should consider socio-economic concerns and size of the population, composition and structure of household, basic and secondary needs of the people to address the problems of unemployment, under-employment, inequalities in income and assets and over-exploitation of natural resources. It is imperative to match people's needs with development and application of alternative set of technologies for optimum utilization of resources and for alternative employment opportunities in farm as well as non-farm sector to solve the problem of vicious circle of poverty and shrinking natural resource base. The main role of science and technology could be to make such necessary intervention, which may blend with the life styles of the people. These interventions must aim at providing simple affordable scientific solutions, which help the individual, save on time and energy for the same life routine, thus creating opportunities for work in the non-traditional areas and augmenting income. Thus, for ST is a necessity for rural transformation by addressing long-term problems of drought, fodder crisis, epidemics, drinking water shortage, health and sanitation, improved dietary habits, nutrition, housing etc., and day to day problems like improved technology for routine work, value addition to the products, shift towards non-conventional energy sources etc.

People's Participation and Empowerment Based on Local Felt Needs

Participatory development implies intensive use of locally available information based on experience and knowledge of the rural people and, in the process, attempts to realize their latent capabilities. These efforts are supplemented with information and resources available from the government and other sources with a view to enhancing local productivity. Local community should be involved from conceptualization to decentralized implementation stage, which will be more effective and fruitful than piecemeal approach. Since traditionally they are directly or indirectly dependent on natural resources, their active participation become crucial to the success of such development and management initiatives. Emphasis should be given to enhance the capacity of people by "learning by doing" and innovating in a changing environment and taking collective decisions and actions by involving all the stakeholders.

Therefore, success in development and dissemination of green technologies lies in participatory systems with down to earth approach for effective percolation of technology from people's acceptance point of view as well as to make proposed intervention sustainable to be managed by them.

Technology, Gender Balanced Development and Empowerment

While focusing on overall rural development through technological intervention, one has to ensure that the ultimate goal is to bring about sustainable development, which should be linked to the carrying capacity based developmental planning by analyzing the underlying co-relationship between gender, poverty and environment. Therefore, measurement of sustainability is the immediate challenge. The visible indicators of sustainability, whether qualitative or quantitative, have also indicated loss of bio-diversity, and degradation of resource base in many areas. According to FAO, the most disadvantaged of all are women, the "silent majority" of the world's poor. A recent study conducted by the Centre of Science and Environment, Delhi in Himalayan village in Chamoli, Uttaranchal unfolds that women spend 59 per cent, children 36 per cent and men 15 per cent work hours every year confirming back-breaking workload on rural womenfolk. These are perhaps, the alarming indicators about the Indian rural women, whose work is filled with drudgery and hardship—leading to feminization of poverty, thus, in real sense excluding them from full participation in developmental efforts to eliminate poverty. Though rural women produce up to 80 per cent of food in the developing countries, yet, studies indicate they have title to only a fraction of farmland, and access to just 10 per cent of credit and 5 per cent of extension advice. There is a growing perception about the link between gender perspective, development and environmental concerns.

Such trends demand innovative approaches, institutional development, awareness and education, and socio-economic motivation for economic development and environmental soundness, which are complimentary to each other. But, such development at village level can be sustainable only when social and economic processes are more equitable, efficient, and endogenous. Need is to weaken the culture of silence and dependency and promote the process of empowerment. Thus, the promotion of gender based equity in access to, and control of productive resources; access to research and extension service benefits; enhancing women's participation in decision-making process at all levels must be pursued in order to address the root causes of persistent poverty and food insecurity among rural women and the families they support. *Here, women's empowerment is operationally concerned with the genuine empowerment through education, ideas, through consciousness, through mobilization and participatory approach to improve their lives in a practical way, which would ensure a sustainable future for rural India. This is where sustainable development must have its roots.* For this, priorities should be on activities which are easily manageable, require low external inputs, contribute to family nutrition, easily marketable, provide regular income and create assets of direct benefit to women. Technology must ensure sufficient production of food, fodder, fuel, energy, clean drinking water supply and fertilizers without spoiling and damaging the environment. Thus, drudgery should be reduced both at household related tasks and occupational tasks. This would energize rural women to move towards active and useful participation in other economic activities and in the overall technological transformation of the society.

Involvement of Voluntary Organizations: Linkage and Networking

In order to facilitate effective rural transformation, it is now being felt strongly that voluntary organizations with scientific and technical expertise can play a crucial role in development and dissemination of technology along with social engineering and motivation work; providing a strong communication link between the users and S&T agencies; maintaining forward and backward linkages; standing as a guarantor/facilitator for loan and support services for the end users and providing marketing linkages, thereby playing a vital role in improving profitability and aptitude to use the technologies. Such organizations often are local in nature or associated with external groups (ST institutions, developmental agencies) have a strong linkages and networks for their collective and mutual benefits. To provide good governance and to reach majority of the people who live in the vast area of rural India, several grass-roots level organizations with scientific and technological capabilities are providing crucial link between the emerging new developments in knowledge and technology and also helping to strengthen and diversify the local economy, utilization of local resources, and to upgrade the skills of artisans, landless labourers and other disadvantaged sections. They can facilitate the articulation of felt needs on one hand, and also match those needs with potential source of technology. It is being felt that such science based field groups can play a crucial and catalytic role in developing field tested and viable models of green technologies for socio-economic up-liftment through skill up-gradation, income generation, drudgery removal, sustainable use of resources, etc. by adopting systems management approach for technology absorption with social and managerial inputs. Such interventions will bring in visible changes in the lifestyles in rural areas and having multiplier effects in different parts of the country.

Systems Approach for Participatory Technology Development and Transfer (PTDT)

For effective rural technology base and rural transformation, there is a need for development and introduction of appropriate technologies coupled with sound delivery system having systems approach for need identification, technology choice and appropriate scaling, technology modulation/innovation, economic and ecological sustainability and optimum use of local resources and material with emphasis on technology capacity building amongst local people. The process of technology development and transfer for direct benefits to weaker sections, women and other backward category of society has some unique features which are quite different from the issues and the processes of the standard method of such transfer mechanisms from research laboratory to industry. *The success of technology development and application oriented programmes of societal nature clearly indicates that more than technology development it is the total system which is important for acceptance by the people leading to creation of sustainable livelihood.*

Therefore, it is essential that various inter-related management components of the technology, namely, "hardware" (materials such as crop variety), "software" (technical know-how, information), "human ware" (human ability), "orgaware" (organizational and management aspects) and the final products (including marketing) are rendered accessible to rural people as consumer/end-users. In fact, a large number of technologies and institutions dedicating themselves for scientific and technological research work are available, but, there is a need for 'systems approach' as management tool to modulate and transfer technology to the field as user and environmental friendly and cost effective so as to have easy adaptability and replicability. Such systems approach involves designing of programme to infuse technical skills in local people and enhance capabilities of artisans/ small and marginal farmers/landless labourers through application of ST inputs for creation of sustainable livelihoods. This bottom-up approach emphasizes development of a complete process of identifying needs, appropriate technology design and development, systems engineering resulting in creation of systems based on sustainable technology packages and maintenance of assets created. The process takes into account the local resources/ skills available, the problems of the existing economic structures, gender and environment friendliness of the technology package and long-term sustainability of the entire programme. *Thus, focus of technology development and transfer would be to a problem solving approach rather than merely developing products or processes for the benefit of rural people and society.*

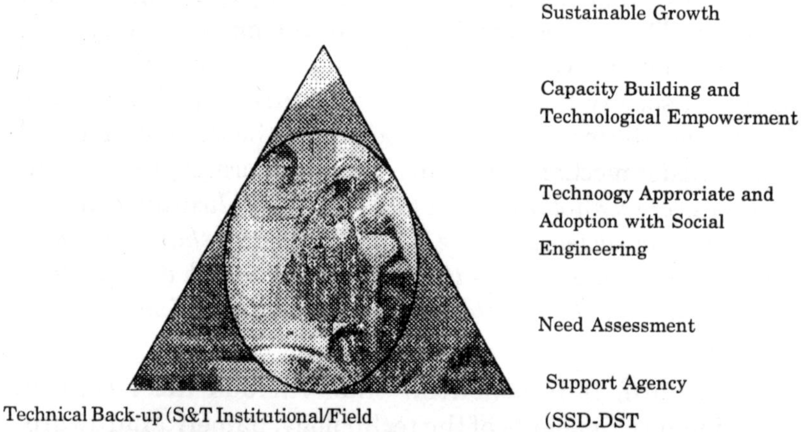

Sustainable Management Model for Technological Empowerment

The salient features could be summarized as under

- The need of technology is assessed from the people with the assistance of local NGOs/ST based field groups usually science and technology field groups. The NGOs concerned then look for an ideal technology option keeping in mind the availability of local resources and skills.
- Technology is developed with their in-house capability or with assistance of nearby technical institution. Thereafter, the technology is appropriated to a scale and level, which

is acceptable to the people, and also which will be suitable in the local conditions for long-term sustainability.

- For a period of 2 to 3 years the people are trained to handle the technology that is being developed and they are also trained in maintenance, service and repair. During this period, technology back up through continuous handholding is provided by the NGO concerned, if necessary, with assistance of the technical institutions.
- The NGOs/S&T based field groups in the meanwhile also sets up the process for backward and forward linkages both for production as well as marketing for long-term sustainability.
- The system once in place with the total package of sourcing of raw material, processing, final product preparation, packaging, marketing, etc. the NGO gradually withdraws giving the entire responsibility to the locally formed people's organization or groups for further dissemination and continuation of activities on self-sustainable basis.

Steps involve in effective technology development and delivery process should, therefore, involve the following as management strategy for better outreach of proposed intervention:

- Need assessment/Problem diagnosis: need identification and technology gaps as perceived by people in the village;
- Identification of technology source (S&T institutions);
- Maximum utilization of local resources which can be sustained economically and ecologically; skills and markets;
- Participatory technology modulation/development which are environmentally compatible and socio-economically sustainable;
- Self-sustaining ST base in rural areas (S&T field groups) to facilitate participatory planning, designing, execution and monitoring of chosen technologies: technology development, field testing, demonstration and training; and finally dissemination to meet the local needs;

- Setting up of viable S&T based micro-enterprise involving ST field groups in rural areas and linking them with national ST institutions/labs for adaptive R&D based on field generated specifications;
- Qualitative shift in production/distribution from household models to collective, networked "industrial" forms for higher productivity, incomes and reduced drudgery;
- Constant interaction between the innovator, the producer and deliverer or technology transfer group for technology assessment and further refinement; and finally;
- Innovations should be in the spheres of both social behaviour and technology to bring about the desired ends of self-reliance, equity and sustainability.

Awareness generation and technology capacity building through demonstration-cum-training among rural people about the importance of technology relating to their needs is also a necessary aspect for technology delivery system. The entire technology package will have to be developed based on local needs, which, in turn must be evolved through a detailed scientific micro-planning exercise. Thus, appropriateness of technology delivery system through S&T base would revolve around effective education and learning process to facilitate the development of creative, innovative and autonomous capabilities in individual and community as a whole through attitudinal/behavioural changes to promote evolution and transfer of environmentally sound technologies.

With systems approach discussed above, appropriate technological interventions can be made at the grassroot level to improve the living conditions of disadvantaged people. Such interventions can brought in visible changes in the lifestyles of local people and can have multiplier effects in other parts through some appropriation as per local needs with technical backup support of its technology generator/provider groups. It will also help in forging partnership for cross-fertilization of ideas and experiences to work towards a socially responsible and environmentally sustainable development in rural areas through available green technological options.

People's participation and capacity building: Essential in planning, execution and management of developmental projects from sustainability point of view.

NGOs and Technology Models

There are several organizations with strong grassroots base active in rural India to represent the poor's interests. They are playing an important role across the rural economy in areas as diverse as animal husbandry, horticulture and food processing etc. to strengthen the linkages amongst the ST institutions, agricultural universities and village people especially, small and marginal farmers having small landholdings. Their role in these sectors becomes crucial as they are ideally placed to innovate, design and disseminate field-tested proven technologies with appropriate modulations which address location-specific problems and concerns on the prevailing natural and socio-economic conditions, needs and priorities. Such organizations with in-house ST capabilities and technical back-up support from nearby ST institutions can play crucial role in helping village community to translate the new technology packages to actual grassroots level activities under demand-driven system. In this process, a number of ST based field groups in different part of the country are actively involved to empower rural community with new skill and knowledge through participatory approach.

The development approach of several of these based on the above systems approach have resulted in sustainable models of activities, where application of science and technology have resulted in improved living conditions as well as new opportunities for gainful employment to rural poor, including women. For instance, in a collaborative programme between the Council for Advancement of People's Action and Rural Technology (CAPART), New Delhi, and the Department of Science and Technology (DST), New Delhi, CAPART has taken up large scale replication of some of the technology models as package of appropriate technologies evolved and proven by the latter through their partners, i.e., S&T field groups along with collaborating S&T institutions through systems approach. In this endeavour, efforts are also being made in mountain regions to make visible impact in improving the living conditions of local people in fragile conditions involving them as "primary stakeholders" at all levels as producer, as consumer and change agent from designing to adoption of cost effective technology package to suit local needs. The development approach of some of these has resulted in sustainable models of activities. Some of them are given below with highlights of the nature of replicable technology models generated in a decentralized manner by adopting systems approach:

Women Technology Park at Dehradun, Uttaranchal: Model with Integrated Approach

The above system approach has been adopted for technology development and transfer in different parts of the mountain regions by some field-based S&T field groups/NGOs involving small and marginal farmers, women, weaker sections, scheduled castes and scheduled tribes. In this direction a group of scientists from S&T field group, HESCO, Chamoli, Uttaranchal working on a project *technology upgradation for fuel/fodder needs* of women in Garhwal hills has been able to train hill women from three selected villages of Chamoli in appropriate techniques for cultivation of forage crops/grasses, silage making, stall feeding and efficient use of nonwoody fuel in monsoon season to mitigate the problem of fodder/fuel scarcity. HESCO has also developed new carrying devices using *Agave fibre,* upgraded harvesting tools and stove, which are more efficient in use, and helping in time saving to hill women who are overburdened by daily chores. Training in construction and management of silo pits for silage making and appropriate storage and preservation of grass/fodder techniques has helped a lot to avoid the hardship faced by women in hills during monsoon season and lean periods. In a recent initiative to address the problem of natural resource management and human resource development, efforts are being made by HESCO to establish a *Women Technology Park (WTP),* a live technology model exclusively for women at Dehradun, Uttaranchal. WTP is providing demonstration and training on need based selected technologies and promoting rural entrepreneurship amongst mountain community specifically to women in and around project area. The park is essentially to show case location specific technologies relevant to women, provide training and also backward and forward linkages so that women would optimally be able to use these technologies to improve their living conditions and minimize hardship faced by them in day to day activities. Main objective of the WTP is to serve as a common platform to document, develop and transfer proven technologies as per women needs in the mountain region. It is also facilitating increased participation of women technology innovators.

Technological empowerment and skill training of people, convergence of income generating activities and creation of sustainable livelihood through the structure by people's institution are major and critical components of the WTP. Other support services like health, counselling, marketing and credit are being taken care of with the support and linkages of other financial/ technical institutions to make proposed intervention sustainable. Some of the technology packages being delivered to women through Women Technology Park are—mushroom cultivation, floriculture, animal husbandry, natural dyes, vanilla bean cultivation, medicinal plant cultivation, nursery raising, fibre utilization, sustainable fodder management, utilization of conventional and non-conventional fruits, organic farming, post-harvest technology, solar energy utilization, processing and weaving of local wool, women's health and information technology.

The concept of WTP is also providing a nucleus of science and technology for technological empowerment of poor people in nearby areas in terms of appropriate technology back-up and management support / services in various sectors like information, organization and planning; watershed management; agro-forestry processing; sustainable agriculture; horticulture and bamboo farming; energy application and environment; and non-farm activities covering simultaneously the social engineering aspects with low cost appropriate technology packages. Such WTP model is serving as a help desk and allowing women to use resources and facilities of WTP appropriately to develop them as individual/ group entrepreneur. Number of women groups from Uttaranchal and Himachal Pradesh have benefited from this WTP model.

Reaching the Unreached: Case of processing and Preservation of Horticultural Produce: Model with Network Approach

Role Model at Village Level—A Success Story

(a) Need Assessment and Conceptualization

Small farmers and other poor growers of fruits, vegetables and other horticultural produce suffer from the pressure to dispose off perishable produce, often at distress prices due to poverty and lack of access to storage or other preservation facilities. India is the second largest producer of fruits and vegetables in the world but processing is very low, approximately 3-5 per cent. It is also estimated that about 20-30 per cent of the total production of fruits and vegetables goes waste due to spoilage at various stages from collection to sale. Thus, not only valuable source of nutrient wasted, but also its producers face impoverishment and consumers have to depend on high cost branded products from organized sector. This is where, Science and Society Division under the Department of Science and Technology, launched a co-ordinated project on Low Cost Processing and Preservation of Horticultural Produce during 1994-95. The programme was initiated with 'systems approach' to set up economically viable and self-sustainable production units having networking of poor/ marginal growers and other weaker sections, especially women through appropriate technology package for value addition at village level involving S&T based developmental field groups and institutions.

One such unit has been established in the Nagwain area, Kullu, HP by Society for Technology and development, a science based field group (based on field survey done to access the need and required S&T intervention), close to the source of raw material and easily accessible to women growers. It was ensured that local market potential should exist and range of raw produce should be available in order to run the units round the year.

(b) Technology Package

The technology package includes some innovative techniques, suitably descaled and adopted equipments/machines to produce conventional products or intermediates like juice, pulp, puree, pickles etc. with a major emphasis on production of innovative dehydrated/osmo-dehydrated products such as onion and garlic powder, dried apricots/plums, apple chips, papaya candy and mango/jackfruit leathers etc. Given the focus on innovative work, and the perspective of adopting low cost energy saving technologies (such as agro-residue based flue-heated driers, zero energy cooling chamber for on farm fruits and vegetable storage based on the principle of evaporative cooling, lactic acid fermentation and other nutrient preservation techniques), major thrust under this programme was given on minimizing pre-and post-harvest losses; increased shelf life with quality control; controlled atmospheric and aseptic packaging, and emphasizing on drying and dried products to have all year round production and sustainable income ensuring maximum benefit to the women groups.

(c) Systems Design: Network and Management

Each location comprises of mother unit at Tehsil level for drying, quality product production, packaging and marketing linked with several home based or cluster of village based sub-units for providing semi-processed produce to mother unit, together covering a cluster of, say, 10-15 villages. Adaptive R&D work, training in quality control and production process is carried out at mother unit, while, overall organization and supervision of the units is done by the S&T field group with a network of women.

Training, an essential element of the programme was designed to be imparted at two levels: at institutional level by resource agency (Indian Agricultural Research Institute, Delhi) to core women group and at field level by trained women core group to network beneficiaries. It was ensured that at every level of operation, network of women is closely involved in production, processing, packaging and marketing promotion.

Empowerment: Mountain Women trained as para-technologist working in horticultural processing unit to make value added quality products at village level

(d) Impact Analysis

As a result of above network efforts, full-scale processing and production unit has been established as common facility centre with technological back-up/training from the Indian Agricultural Research Institute (IARI), Delhi. The project was co-ordinated by the Centre for Technology and Development, S&T based field group based in Delhi to provide technical back-up as well as managerial support to participating partners in this project. STD has carefully developed low cost technology package, recipes and processing protocols through field trials based on traditional knowledge/techniques upgraded with modern ST inputs, availability of raw material in the area and market demand of innovative products. Technologies of appropriate scale/type were introduced or wherever necessary adapted/standardized under field conditions with functional networking between different levels of production. This Unit is having a core team of 4-5 trained women with online quality control lab. facility. These trained women as para-technologists are now involved in imparting training and technical know-how to other women of the area as ongoing activity having multiplier effects. Unit also has got requisite FPO license and is now self-sustainable, being managed by women co-operative to process, preserve, pack and market the finished products. These innovative products are now being marketed successfully by them under the common brand name "Farmer's" ensuring maximum benefits to women growers/beneficiaries at the village level with an approach of value addition at source itself— *thus reducing dependency on middle-men, reduction in post-harvest loses, saving in transportation cost, and time.* A new sense of accomplishment by way of new skills and earning independent incomes have given them the confidence to bring about other reforms in their lives. Many of them are now developing fruit belts around the unit through better nursery plantations techniques for better production. It is estimated that about 70-80 families around this unit have benefited directly to have all year round production for self-consumption and sustained income with minimum wastage. Many of innovative products: natural plum and apricot drinks, rhododendron squash etc. made from farm fresh produce by the caring and knowing hands of trained women are now available in the market. Each unit is estimated to have year-round production of 500 kg/day of finished products with monthly turnover of Rs. 50,000 to Rs. one lakh. *Moreover, local processing has helped not only in extending the shelf-life of otherwise perishable produce, but also in generating local employment in rural production system.*

(e) Success and Spread Effect

Success of this technology model lies in "bottom up approach" and linkages amongst DST, S&T based Field Groups, ST Institutions and above all target group (Women) — from planning to implementation stage as partners, carriers and change agents in technology development and transfer in decentralized way of functioning. This has resulted in long-term sustainability by linking the women from small grower families in the area with the production and marketing system-involving their full participation at all levels, thus, making them self-confident. Now, all units with trained women are serving as resource agencies to replicate this model in other areas in respective states.

As a village based S&T intervention, it is well suited to rural women as income generating activity based on sustainable use of local resources and material for large scale replication in rural areas. Seeing the success potential and 'systems approach' in planning and implementation of the above programme, recently a UN agency—Asia Pacific Centre for Transfer of Technology, Delhi as well as Women and Child Department have taken the initiative in collaboration with DST, Delhi to replicate this technology model for large scale replication in the North Eastern States and other areas, and is therefore, a path-breaking effort to carry S&T for the socio-economic upliftment of poor rural women.

Photographs showing

1. *Geranium* cultivation in polyhouse condition as cash crop

2. Discussion with women groups

3. Discussion during visit to pottery unit

4. Visit to see construction of water channel to meet irrigatineeds- under various projects being implemented by field

On similar decentralized approach in mind, STD has also made fruitful contribution to strengthen and revive traditional artisan and their trades in Mandi and Kulu area of Himachal Pradesh with appropriate technology packages to enable artisans and local growers to compete with organized system as reflected in Table 1.

Table 1
Impact of Decentralized Technology Packages for Traditional Artisans and Growers: Case of STD, Mandi

Projects and Area of Intervention	*S&T Component and Technology Package*	*Impact: Empowerment and Benefit*
Lather Tanning and Carcass Utilization	• Introduction of better processes and equipments linked to network of flayers • Use of pressure vessel to cook carcass of fallen animals • Technical backup from CLRI, Chennai	• Ensures production of better quality tanned leather with network of traditional flayers (170) • Qualitative shift from household level to collective form of production for high productivity and enhanced income • Diversified products: wallets and purses etc. • Products such as bone meal and meat meal as fertilizer and animal feed • Environmentally friendly technology
Rural Red Clay Pottery	• Improve clay forming, processing and firing methods by using additives • Simple de-scaled machines, better kiln and low cost mechanical wheel • Technical backup from CGCRI, Kolkata	• Enabling traditional potters to utilize local clay to make quality glazed tableware, kitchenware and novelty items • Systems provided economies of scale and substantially higher productivity in organized way (10-15 potters) not available to individual potter

Projects and Area of Intervention	*S&T component and Technology Package*	*Impact: Empowerment and Benefit*
Low cost Construction Technologies	• Stabilized mud block making • Micro concrete roofing tiles using local material	• Local masons trained in production technology • Promoting entrepreneur activity
Water, Health and Sanitation	• Introduction of ferro cement latrine; water harvesting structures; awareness about cleanliness and hygiene	• Better sanitation; low disease spread • Water channel and other structures being managed by community
Horticultural Processing	• Establishment of economically viable rural unit using descaled and adopted equipments • Network of small growers • Technical support from IARI, Delhi	• Unit functioning with FPO license • Innovative and value added products • Reduction in wastage loss; elimination of middle man • Systems provided economies of scale and substantially higher productivity in organized way (with a network of 60-70 women growers) not available to individual grower.

Local Capacity Building: The Case of Mushroom and Vermi Technology Transfer in Hill Area: Household Level Model

Mushroom cultivation technology offers a very good alternative to the conversion of organic waste material to palatable proteins. In an innovative approach, Himalayan Research Group (HRG), a field based NGO of scientists in Shimla is working on transferring this technology for the socio-economic upliftment of rural people in the Himalayan region. HRG is using cowdung and locally available agro-residues for producing mushroom compost, which is easily available in hilly areas in required quantity. Intervention is being popularized in Chachyot Block of District Mandi, in HP. HRG has been able to organize poor people especially women in small self help groups of 10-15 members. Each self help group is imparted intensive training on group management and mushroom cultivation (Agaricus bisporus) technology as income generating programme with technical support from National Research Centre for Mushroom, Chambaghat, Solan. HRG has experienced that popularization of mushroom cultivation technology among rural people is mainly constrained by the non-availability of the raw material (compost and spawn) and technical support required from time to time. Therefore, it has understood the problem at the grassroot level and decided to set up a 20 ton capacity mushroom compost unit. HRG is providing preseeded pasteurized compost (spawned) to the organised self help groups along with technical backup to the growers to sustain their units. HRG has developed effective linkages for the marketing of the grower's produce through District Rural Development agency (DRDA) Mandi and Horticulture Produce and Marketing Corporation (HPMC), H.P. in addition to the open marketing to provide maximum benefits to the growers. The Group is also exploring the possibilities of the biotechnological studies in the near future on the testing of local substrates for cultivation and production of high yielding and disease resistant varieties of other mushrooms in that area.

The Himalayan Research Group, Shimla has also been able to organize and train about 170 poor women in Dhaingarya village, of Chahiyot block of Mandi to produce mixed compost, vermi-compost, fortified compost and bio-pesticides as well. These trained women after number of agronomical trials and standardization workdone by HRG, are now able to manage such activities in groups from production to making finished quality products. HRG has helped them in organizing micro-credit to construct sheds for compost production, which is now being sold under brand name. These long neglected hill women follk, are now self-reliant to market these products on their own through a Federation formed to oversee and co-ordinate various operations of a number of functioning units. HRG, which has been able to get rolling this technology on the ground is responsible for entire operational flow from production to inventory to markets ensuring forward and backward linkages.

Women engaged in mushroom cultivation: Household Level Technology for Income Generation

Impact Analysis

Therefore, such innovative attempts by HESCO, STD and HRG show NGOs potential in developing and optimizing appropriate technology packages applicable at household as well as in organized way involving cluster of villages not only from the technology point of view but also with good management and governance practices. In each model efforts have been made to incorporate simple technologies as generic/unit technologies for cost effective output but also in generating job opportunities through self-employment at village level. Indirectly, they are also helping in promoting related income generating activities like floriculture, cultivation of important medicinal and high value crops like T. bacatta, geranium, scented rose and forestry related activities in villages . Their approach also justify 'problem based intervention' rather than 'solution based thinking' with systems approach covering managerial and social engineering aspects as well. It also justify that related technology package was created with technical back up support from nearby S&T institution and establishing marketing linkages in response of the needs of the rural poor, who have little purchasing power. In all of above focus was also given on involving landless labourer/ poor farmers/women growers as 'primary stakeholders' at all level as producer, as consumer and change agent from designing to adoption of cost effective technology package to suit local needs.

Such an intervention as in the case of horticulture and tissue culture nursery raising also assures an increase in the net economic return, compared to that of the conventional approach/method, for a certain unit of land, water, time and capital which is very important from the farmers point of view. Some other potential areas like cultivation of off-seasonal vegetables, need-based cultivation of herbs for local and community utility, technology based micro-watershed management, preservation and processing of horticultural produce with quality control have shown promises for sustainable utilization of local resources in hills and better livelihood opportunities to local people in mountain production system. If proper networking amongst different stakeholders in terms of technology, credit, and social engineering based on location specific need is ensured from planning to implementation stage, it certainly helps in percolation of technology at the grassroot level and post-project sustainability.

Above experiences clearly indicate that the "bottom up" approach for participatory technology development by adopting "triangle model" starting with the people and with lateral contribution from Science and Technology Institutions and close interface with voluntary organizations can provide sustainable models for technology development and transfer. A new sense of accomplishment by way of new skills and earning independent incomes generate confidence amongst people to bring about other reforms in their lives. Application of systems approach with appropriate networking and linkages amongst people and institutions as described above can certainly change the lives of the people through technological empowerment and capacity building at the grassroots level putting in place sustainable systems, which are managed, by the people. The results are providing clear pictures of changes and evaluate the effect of appropriate technological application in individual as well as integrated and holistic development and management workdone by voluntary groups for enhancing sustainable livelihoods. In such technology models, local problems were addressed to generate following assets essential for enhancing sustainable livelihoods:

1. **Natural Assets:** For this, efforts were made for above and below groundwater rain water harvesting, storage and its judicious usage; channels; production plantations of various types like fodder, forages and grasses to fetch quick returns initially and timber for long-term benefits ensuring immediate community participation. Ex. Water, Health and Sanitation Project at Mandi.
2. **Physical Assets:** Mechanical check dams; drainage work; gully plugging; agro-processing centre; production-cum-demonstration training centres for bio-compost; community centres; and pucca roads etc. Ex. STD and HRG, Mandi.
3. **Human Assets:** New employment opportunities with specialized skill and knowledge about processes, quality production, value addition and marketing in the area of bio-dynamic gardening, organized compost making, nursery raising, tiles making, glazed pottery items and trading activities etc. have generated immense confidence amongst local people and self help groups due to enhanced income level, reduced drudgery and additional off farm employment leading to sustainable growth.
4. **Financial Assets:** Above interventions were found to be economically viable with direct employment and enhanced income in new trades; horti-processing; leather tanning, etc.
5. **Social Assets:** Formation of water management committee, Women's SHGs; monetised (thrift, informal and infrastructure bank) and non-monetised (biomass, service and labour bank) banking structure has helped in increased participation and collective decision ensuring social security.

Conclusion

Technology models and case studies described above clearly illustrates how scientific and technical interventions can improve the quality of life in rural areas. Details findings of above models based on field visits in respect to management approach, technology choice, linkages, role of NGO and institution, and weakness/issues for better outreach is reported in particularly in the context of rural mountain areas. The experience, has brought to light the fact that there is vast potential, intelligence, strength and ability in our rural people that can be tapped and used for sustainable development. For ensuring the involvement of people, structures were created for internal resource generation by formation of new groups, village level organizations, apex committee at the watershed level; strengthening and stabilizing the structure through various inputs in the form of training, creating work opportunity alongwith social and environmental security; and introduction of informal system of non-monetized and monetized banking structure managed and controlled by villagers linked to the Panchayati Raj institutions for further continuance and sustainability. Such diversified livelihood options have resulted in negligible migration of the people, stability and social cohesiveness. In the success of these efforts, the guiding principle has been to provide enough flexibility and space to help the people and groups to come out with their own plan in technological and societal areas like eco-regeneration, etc. This factor has contributed in significant way enabling people to shed off the inhibition and doubts and share the responsibilities together. The people are able to access the technical, social and economical consequences of their management decisions. Now, with the convergence of various programmes and technical backup support by nearby institutions over the last couple of years, the micro-planning system of each project is well strengthened and entirely owned by the community organizations.

Above case studies also show that following factors/ approaches are crucial for empowerment of people as well as to ensure sustainability of developmental programme for effective rural transformation:

- To deal with rural poverty, need is to deal with ecological poverty through participatory approach for ecological restoration. Gender integrative participatory technology development/appropriation.

- Eco-restoration and employment generation through judicious use and transfer of appropriate technology package appropriated with involvement of people at all levels both as "change agent" as well as "consumer" of technologies—not as beneficiaries.
- Gender-sensitive planning at the local Panchayat level with gender/sex-segregated information to support S&T based developmental efforts in rural production system.
- Proper recognition of the productive and domestic roles of women.
- To make such intervention successful, other areas of interest and support activities to provide basic facilities like drinking water, health and sanitation, nutrition, family planning, adult education and social security etc. should be taken care for overall sustainable community development to have a catalytic effect. With such support services, access to training and education or participation in economic and social life become a reality for rural poor.
- Importance of S&T Base (Science Based NGOs) in the Actual Area for effective technological intervention/ to develop location specific technology package taking into account the perceptions of people.
- Involvement of local motivator as active "change agent" or "natural carrier" of technology through dove-tailed approach of motivation through talks and audio-visuals, awareness build up through demonstration, and hands on the job training in relevant field.
- Strong institutional linkages between S&T field groups, R&D institutions with an extension machinery to disseminate the proven technological package at grassroots which empowers and enable rural people to seek local solutions and have face-to-face interaction between them and scientists for proper scientific know-how, and
- Equal sharing of benefits amongst different stakeholders.

Therefore, participatory technological development and dissemination is a necessary process that has to be set in motion to develop sustainable technology models for replication which will have an impact not only on the economic sphere, but also on the social, political, cultural, moral and environmental levels in rural production system. As in rural areas, the entire activities go around womenfolk and women's active participation should be ensured at every level from planning to implementation stage both as change agents as well as beneficiaries for the achievements of goals of equality and development and empowerment by changing social attitudes and elimination of all forms of gender-based discrimination — will be a guiding factors for any future development planning. Further, in the mountain societies, women are responsible for most of the farming as well as the nurturing of future generations. A central focus of ST based mountain development initiatives should also be to decrease these inequities, aiming at gender-balanced development requiring actions from all of us.

The concept of decentralized technologies, local resource use both conventional and non-conventional, strengthening of local market and upgradation of human skill have been largely ignored in mountain parts of the country. Strategies of integrated approach have although been mobilized but appropriate technology packages were simply overlooked causing major failure of the approach. This also gives us the understanding that technology transfer is not the answer, but technology adoption at various levels is needed for desired results.

Hence, in a country like India having about 5,91,000 villages, each of them covered by one or other programmes of rural development, the need of the hour is to devise strong and well-knit technology delivery systems involving science based committed voluntary organizations in identified areas and dovetail them with such programmes. This requires a systems approach regarding needs—identification, choice of technology, appropriate scaling, technology upgradation/innovation, economical and ecological sustainability and optimum use of local resources with emphasis on technology literacy amongst rural community through mass communication for upgradation of skills and capabilities to address most important issue of human resource development and natural resource management in rural India.

Suggested Readings

Agarwal, Bina. (1995): Feminist Environmentalism or Eco-feminism. *The Hindu, Survey of Environment,* pp. 1-14.

Agarwal, Bina. (1997): Gender, Environment, and Poverty Inter-links: Regional Variations and Temporal Shifts in Rural India, 1971-91, *J. World Development,* 25(1): 23-52.

Agarwal, S. and Kumar K. (1993): Science, Technology and Women for Sustainable Environment and Rural Development. *Int. J. Sust. Dev.* 1(3): 56-60.

Department of Science and Technology, Government of India (1995): Technology Models for Rural Applications, *DST Publ.,* N. Delhi, p. 115.

ESCAP (1997): Women in India. A Country Profile. United Nations, New York.

H. Bongarta and D.R. Dahal, Development Studies: Self-Help Organizations, NGOs and Civil Societies (Kathmandu, Nepal Foundation for Advanced Studies, 1996).

Khosla, A. (1991): Role of Science and Technology in Sustainable National Development. *In Proc. National Summit on Technologies for Environmentally Sound Sustainable Development,* Development Alternatives, N. Delhi, India, pp. 4-7.

Kwaschik, R., Singh R.B. and Paroda R.S. (1994): Technology Assessment and Transfer for Sustainable Agriculture and Rural Development in the Asia-Pacific region: A Research Management Perspective. (Research and Technology Development Service, SD Department, FAO). M. Flaherty, P. Vandergeest and P. Miller (1999): Rice Paddy or Shrimp Pond: Tough Decision in Rural Thailand, *World Development,* 27(12).

Papa R. and Shanmugasundaram Y. (1996): Proc. Women and Emerging Technologies (lndo-British Seminar), British Council, Chennai. p. 255.

Raghunandan, D. (1988): Rural Development through Science and Technology Application, *Kurukshetra,* XXXVII(1), pp. 77-81.

Raj Rani. (1997): Environmental Degradation and Women, *Kurukshetra,* XXLV (4 and 5), 57-59.

Schumacher, E.F. 1973: *Small is Beautiful*. Harver and Row, New York.

Sharma, M. (1991): Science and Technology for Rural Development. *Yojana,* 35 (1 and 2), pp. 78-83.

Time Newsletter, DST Publication, Vol. III, pp.1-22, October 2002.

Venkateswaran, S. (1992): Living on the Edge: Women, Environment and Development, Friedrich Ebert Stiftung, New Delhi.

UNDP (1997): *Human Development Report,* Oxford University Press, New York.

World Bank (1991): A World Bank Country Report: Gender and Poverty in India, World Bank, Washington, D.C.

World Resource Institute, World Resources, 1998-99, Chapter 9.

Notes

- Present paper is based on the experiences gained in handling ST based developmental projects by the authors and field visits made to project sites. Authors gratefully acknowledge the inputs provided by field agencies and support provided by DST, Delhi for this work.
- Views expressed in this article are those of the author, not of the organization to which they belong.

7

Technological Nurseries for Rural Industrialisation and Entrepreneurship Development

Ram Prasad

Introduction

There is a deep crisis of credibility of science in development. The industrial research establishments of corporate world are working at an amazing innovation pace. The shadow chasing scientists of developing world are caught in the whirlwind of chaos and confusion created by innovative wits of the corporate world. We are not equipped to do R&D at the same pace as the industrial research laboratories of the corporate world can do. But our problems are our own and we know them better than the corporations of the developed world. We can certainly apply our scientific infrastructure to our problems in accordance with the spirit of the 1958 Scientific Policy Resolution which says, "The key to national prosperity, apart from the spirit of the people, lies, in the modern age, in the effective combination of three factors, technology, raw materials and capital of which the first perhaps is the most important, since the creation and adoption of new scientific techniques can, in fact, make up for a deficiency in natural resources, and reduce the demands on capital. But technology can only grow out of study of science and its applications".

This crisis of credibility is not a new phenomenon. It started soon after the national laboratories and research institutes became functional. The scientific resolution of 1958 was a critical, sensitive and conditional document and it was not properly understood by many. Jawaharlal Nehru was himself concerned the way these laboratories and research institutes were functioning. It is on his suggestion that the Symposium on Science and the Nation during the Third Five Year Plan (July 27-30, 1964) was held. The science leadership of that time was not interested in taking the responsibility of implementing the depth of the recommendations of this symposium nor were they interested in the depth of the Scientific Policy Resolution. The concept of technological nursery emerged from the thought process generated at this 1964 Symposium as a virtual industrial laboratory of the science driven industry that was yet to be born. This chapter discusses concepts, strategy and perspective of science driven rural industrialisation facilitated by these technological nurseries.

Confidence as Resource for Change

The crisis of credibility in science and technology is the result of the crisis of confidence—a psychological factor. This lack of confidence in creating and adopting new scientific techniques to make up for a deficiency in natural resources, and reduce the demands on capital arises from what Jawaharlal Nehru defined as the lack of scientific temper. This is an attitudinal problem which could be corrected by changing the attitude — to look at things as they are and not as they are believed to be. It was not Jawaharlal Nehru's intention to ridicule religious and customary obscurantism as many of his admirers interpreted. In his last will and testament he said: "I am proud of that great inheritance that has been, and is, ours, and I am conscious that I too, like all of us, am a link in that unbroken chain which goes back to the dawn of history in the immemorial past of India. That chain I would not break for I treasure and seek inspiration from it." Clearly his instance was not anti-religion or anti-past. Scientific temper was the attitude that looks at things as they are and not as they are believed to be.

It is not easy to change the old attitudes built in individual's character. Jawaharlal Nehru recognised, "And though I have discarded much of past traditions and custom, and am anxious that India should rid herself of all shackles that bind and constrains her people, and suppress vast number of them, and prevent the free development of the body and the spirit." In his prevent the free development of the body and the spirit." In his inaugural address at the Science Congress in 1957 at Calcutta he said, "I plead with the scientists here and elsewhere to remember that the scientific spirit is essentially one of tolerance, one of humility, one of realisation that somebody else may also have a bit of the truth. Scientists should note that they do not have monopoly of the truth: that nobody has monopoly, no country, no people, no book. Truth is too vast to be contained in the minds of human beings or in the books however raced."

The reality is that while it is essential to have a scientific temper, most of the people, almost all, fall short of this attitude to varying degree. Individuals are born as individuals—they carry inward-looking attitudes of natural being and not the outward-looking positive social attitudes that strengthen society and route benefits to individuals in the long run. These are to be cultivated — just as Nehru had tried to cultivate them. The attitudes of the prevalent environment are naturally cultivated in the individual. The social system tries to maintain status quo—thanks to these built-in attitudes. Development on the other hand is change-oriented. Science works as the driving force towards change. Scientific temper helps to keep the direction of change focused towards better things. This provides individuals and the society to gain confidence and this confidence becomes a valuable resource.

Elementary Industrial Laboratories

But scientific temper alone cannot help to bring about a change. The process of change requires an organisation to foresee the direction of change, to define the destination, to build the route and to carry the social system along it to the desired destination. This organisation itself is a social system comprising science as path finder and technology as the vehicle. Now technology development too is an organisation and this organisation exists, in fully developed form, in the shape of an industrial research laboratories of the corporate world. The national laboratories and research institutes work to produce research products in natural, applied and social sciences—they are no substitutes for industrial research laboratories of the corporate world. Their products are fruits for the consumption of the society whether they are eatable or usable or not. In developing countries technological nurseries are require to work as elementary industrial research laboratories more virtual than real. Nehru organised national laboratories and research institutes to carry out RD in different areas of natural, applied and social sciences. Technological nurseries are required to use these laboratories and institutes and work as technology vehicles for the developmental change. There is stiff resistance. Why?

To understand this point more clearly we may make a reference to the book, *Innovation and Change in Organisations* by Nigel King and Neil Anderson (1995, Routledge, London and New York). This book provides a critical introduction to the psychology of innovation and change at work — a broad, complex subject, and one which crosses many of the traditional boundaries within psychology. "Resistance to planned organisational change has long been an issue of concern for organisation development practitioners and management scientists. Resistance has been seen at best as disruptive and troublesome and at worst as a co-ordinated process of radical militancy designed to undermine the very fabric of managerial control. But such views are revealing of the dominant perspective in resistance to change research and organisation development practice".

Structural Resistance as a Resource

In developing countries, however, it is the system governance that is opposed to real change and thereby to development. The leaderships of the parties in opposition and even trade unions are also with them to oppose the real change. Only a minority of enlightened individuals —visionaries — are supporter of change and development. This situation completely turns upside down the belief of the developed societies that resistance has almost always been characterised as irrational, counter-productive behaviour engaged in by a minority of workers to the inevitable detriment of the organisation, and in the long term, to the disbenefit of those employees themselves. It is the majority of people in the system governance that works against themselves in the long run and not the minority of visionaries or vast population of indifferent and unconcerned people who constitute the nation. "The point is that 'resistance' is wholly in the eye of the beholder. It is entirely possible that the actions of an individual within an organisation may be perceived as harmful resistance warranting summary dismissal by some, and simultaneously as morally justified and heroic behaviour worthy of the highest praise by others." Who is right or wrong on either side of these fundamentally divergent views and attitudes is determined by the state of the society in which we find ourselves in future. For determination of this state in advance we have to take resort to holistic research and development.

Structural resistance provides us with a powerful resource in holistic research and development. The outcome is first defined as virtual reality—the reality that does not exist in reality but visualised as a virtual being. This virtual reality cannot materialised in its total volume - but it can be realised as a real reality in a nominal volume to begin with. The current life style in the West, which is a real reality for the people who enjoy it, may be regarded as virtual reality for the whole world as an elementary assumption. This lifestyle is dependent on the market economy of the societies that enjoy it. These societies are equipped with production facilities for goods and services for their own consumption and consumption of the rest of the world. But they find sustainability extremely difficult. Organised marketing is a equally big problem for all industries from the tiniest to the most sophisticated ones. Those who make investments in holistic research and development to link production with changing social needs and the changing competitive environment are able to maintain sustained growth. But this sustainability requires huge R&D investments which only a few in the corporate world can afford — there is really no lasting sustainability consumption centred activity. The structural resistance by Nature to industrial onslaught by humankind has placed most sophisticated industrial societies in a position of humility. They know that the societies which are considered most backward are reality not so. They can reach the stage of sustainable development more easily than the industrialised countries.

Holism of Development

But holism requires much more — it must include the sense of all inclusive totality. Interestingly this emphasis on all inclusive totality places at the disposal of less developed societies, an inexpensive volume of relevant R&D at the highest level of the corporate world which is struggling hard on the same problem of development — how to change the ongoing in order to survive in the world of future. Why? Complex dynamic systems, whether physical, biological or societal, have the same underlying internal dynamics and exhibit common emerging patterns. All that is needed is the cultivation and development of re-investment skills.

In conclusion the book *Innovation and Change in Organisations* admits that resistance to change remains a fascinating but surprisingly under researched topic in organisational psychology. At first glance resistance behaviour appears to be a rather simple, unidimensional action, but the more deeply one looks into this topic the more complex, multi-dimensional and value laden one's perceptions seem to become. Early management research in this area was dominated by the desire to identify and overcome resisting forces, and this theme has prevalent to the present day in the managerial literature. We have argued at several points in this chapter that such a perspective has unjustifiably oversimplified resistance to change by portraying any form of resistance as harmful, illogical and counter-productive to the mutual benefit of managers and employees alike. But a careful examination of the reasons why resistance occurs in the first instance and a review of its many possible forms of manifestation on organisations reveal that resistance can often be wholly justifiable behaviour under the circumstances. Moreover we have argued that resistance can, on specific occasions, be a highly productive counter-balance against poorly planned organisation changes imposed from above.

Structural resistance to change can be studied from four distinctly different perspectives —resistance as natural response, as class struggle, as constructive counter-balance, and as individual and organizational restructuring. "More research is needed under all these four perspectives to redress the imbalance of the existing literature with its preponderance of managerialist research in the 'how to overcome resistance' vein. We must move away from seeing organisational members who resist change only as bad patients who refuse to take their medicine; resistance may equally be seen as a defence mechanism against pathogens of change". This is where the book *Innovation and Change in Organisations* leaves us. It is the dead end of the research which the prevalent mechanistic environment of the West allows. For further research we have to come to the organic environment of the developing world.

Testing the Concept

In the introduction of this chapter we mentioned that the concept of technological nurseries emerged from the 1958 Scientific Policy Resolution (SPR). Technological Nursery for Optics Research and Development (TNORD)—the pioneering technological nursery—is an experiment to demonstrate how a full and complete implementation of the 1958 Scientific Policy Resolution (SPR) is possible using its preamble as the guideline. Thus this exercise is an extension of the most gigantic experiment of development in human history that is the above SPR and it too stands alone. An original idea is not well defined to begin with and it needs a host to work as a champion and develop it into an acceptable form. The process of development remains a lonely affair during this phase. This is the test of an original idea. After the completion of the first (lonely) phase the process of development is ready for multiplication.

The chapter then focused on the first psychological factor—confidence as a resource for change, the organisational factor—elementary industrial laboratory (technological nursery) as a tool of change, second psychological factor— structural resistance as a barrier to change as the and the totality factor—holism of development as the goal of the change. The technological nursery as a tool to tackle structural resistance inspires confidence. This confidence is necessary to realise the planned change. The scientific infrastructure came into being because of the confidence of Jawaharlal Nehru. It survived because of the momentum given to it by the confidence of Jawaharlal Nehru. Nehru was concerned when he found that the scientific infrastructure was not playing the desired role. The concept of technological nurseries was not there at that time and structural resistance was considered as the only reason for it.

Of course the community development programme and the Khadi and Village Industries Commission were already there when the Scientific Policy Resolution was passed. It was felt that these organisations had fallen victim of the structural resistance and lack of scientific temper in the country. It was hoped that presence of national laboratories and research institutes would take care of these activities. The missing link—the concept of technologsical nurseries emerged only from the 1964 Science and the Nation Symposium and it took some time for the idea to mature into a practical shape in the form of TNORD. By that time Nehru was no more and the political responsiveness of the government to new ideas started diminishing although the government continued to provide funds to science liberally.

Like the Indian outlook as represented in the preamble of SPR, TNORD today may be a weak model, partly wasteful and at times agonisingly slow. But let us remember that the Western progress has reached its wasteful end and it seeks support of the rest of the world for sustainable development. The Indian experiment has acquired a new status now. "Success or failure of Indian experiment will make a turning point in world history. Stable human societies based on willing cooperation of the people and their maximum happiness are no easy matter to achieve... Let us pledge that it is our task to overcome the weaknesses of the Indian World Outlook because to us there appears to be no other way possible either for India or for humanity" said the Association of Scientific Workers of India (ASWI) as a preface to the Science and the Nation symposium in 1964.

Focus on TNORD

The background of TNORD is thus the history of the country's hunt for a science driven paradigm to bring about a realistic prosperity for the vast Indian population. The country had a colonial past but this curse was also accompanied by the glory of the freedom struggle and both these realities place before the country valuable resources provided we take a positive assertive view towards development. The inherited colonial machinery opposed development and several developmental activities had to survive by working in struggle made in the given social environments and TNORD is a specific case example of such activities. The history of TNORD is a witness to this process as successive Annual Reports of TNORD discuss. Each annual report of TNORD is a sample record of what happened in the country in the sphere of fostering science driven industrialization process during the year of report. To overcome the limitations of the local environment in which an activity works it is necessary to weigh its experiences in the global context. The eighth report used the book *The Management of Innovation* by Tom Burns and G. M. Stalker, Oxford University Press 1994 as a tool for unshackling the colonial grip and found that even in the global context the process of development needs an organic form of organization rather than a mechanistic form.

The relationship of TNORD to the 1958 Scientific Policy Resolution (SPR), the relationship of SPR to the scientific workers' movement (started in 1918) and the Social Function of Science (1938), the relationship of scientific workers' movement to the struggle of Royal Society makes gives this tiny being a depth that is not well understood by the scientific community. Similarly the relationship of TNORD to the Optical Society of India and to Dr. C. S. Rao, the relationship of Dr. C. S. Rao to Dr. I. R. Rao, the relationship of Dr. I. R. Rao to Sir C. V. Raman and the relationship of Sir C. V. Raman to Optics and Indian science gives TNORD an academic depth. The relationship of TNORD to Dr. (Maharishi) K. Arunachalam, Dr. P. K. Kichlu, Mohan Parikh, Dr. Devendra Kumar and other Gandhian personalities gives the organization a depth along the social axis. This makes TNORD a deep planned enterprise. What is deep planning or depth planning of enterprise? We may go into the detailed project report of TNORD in a greater detail under the next sub-title in a sketchy (information presentation) form.

TNORD as a Deep Planned Enterprise

Technological Nursery for Optics Research and Development (TNORD) came into the being on the recommendation of a week-long Seminar-cum-Workshop on Technological Nursery for Optics Research and Development (NPL, New Delhi, Nov. 19-25, 1988) organised by the National Physical Laboratory. Professor K. J.

Rosenbruch, Chairman on the Panel on Optical Instruments of the International Standards Organisation, had been specially invited to help in the planning and implementation of this programme inaugurated by the Chief Minister, UP.

The detailed Project Report of Technological Nursery for Optics Research and Development envisaged setting of technological nursery for optics research and development in UP hills as a joint project of NPL (CSIR) and the Hill Development Department (UP Government). It was to be implemented as a pilot plant of NPL in UP hills in collaboration with the Garhwal Mandal Vikas Nigam. The seminar-cum-workshop resolved to constitute a body of authorities concerned with the programme on the pattern of CSIR and INSA. NPL accordingly prepared the memorandum of association of TNORD and got it registered under Societies Act of 1860. On the formation of TNORD the NPL activities on design and development of optical systems were transferred live to the new body. These activities form the system vision laboratory of TNORD.

Programme Enrichment at the Seminar-cum-Workshop

In the Seminar-cum-Workshop on Optics Research and Development, besides the delegates from the entire optical community, delegates from the concerned department of the UP government and concerned agencies in UP hills participated. The whole programme was divided in 12 Sessions. Each session was, in fact, an independent seminar and the chapter of the Detailed Project Report were taken as the topics for discussion in most of these sessions. Other sessions were devoted to the associated socio-economical topics. Table 1 lists these sessions

Table 7.1
Session Seminars of the S0eminar-cum-Workshop

SESSION - 1	*SESSION - 2*	*SESSION - 3*	*SESSION - 4*
Social Science Aspects	Conceptual Aspects Inaugural Function	Management Aspects	Technological Aspects
SESSION - 5	*SESSION - 6*	*SESSION - 7*	*SESSION - 8*
Machinery & Materials	TNORD Programme Design	Current Trends In Optics	Optical Testing Facilities
SESSION - 9	*SESSION - 10*	*SESSION - 11*	*SESSION - 12*
Kichlu Centre	Environment & Energy Options	Marketing of Production	Future Planning

Each session critically reviewed the contents of the concerned chapter of the Detailed Project Report. It was felt that the System Director of TNORD would straight-away undertake research work in each of the above directions.

The Force of Enthusiasm

The above table shows the width of the programme and participation. The following (Table 2) gives the depth of the background support from the sponsors and the continued post-seminar activities.

Table 7.2
Cosponsors and Publications of the Seminar-cum-Workshop

	Cosponsors		*Publications*
1	National Physical Laboratory	1	Seminar vision
2	Optical Society of India, Calcutta	2	Seminar Document
3	DRDO, New Delhi	3	Strategy Document
4	IIT, Delhi	4	Social Science Dimension
5	ISRO, Bangalore	5	Background circular
6	CSIO, Chandigarh	6	High tech for countryside
7	CSIR, New Delhi	7	Employment and Environment Link
8	DNES, New Delhi	8	Utilising ethnic resources
9	DST, New Delhi	9	Development oriented micro-climate
10	CST-UP, Lucknow	10	Whirlwinding of rural industrialisation
11	UGC, New Delhi	11	Resources for developmental whirlwind
12	NISTADS, New Delhi	12	Site seminar document
13	INSA, New Delhi		
14	ASWI, New Delhi		

(Cont...)

(contd. Table 2)

	Cosponsors	*Publications*
15	SISI, Okhla, New Delhi	
16	IARI, New Delhi	
17	CRRID, Chandigarh	
18	CSDI, Bombay	
19	CDC, New Delhi	
20	IMMI, New Delhi	
21	ISI, Bangalore	
22	NPC, New Delhi	
23	PEF, New Delhi	
24	Wissitex, Bombay	

Pert-Chart and Milestones

The final session of the Seminar-cum-Workshop considered the CSIR Action Plan and considered programme of project implementation step by step and endorsed it. Operational details are given in the following table.

Table 7.3
Operational Details of Implementation of the Programme

Main Task	*Details*
Organisation	Broad acceptance of Detailed Project Report
	Creation of Joint Implementation Machinery for the Project
	• Steering Committee
	• Memorandum of Association
	• Constitution of the Joint Body
	• Registration of the Joint Body
Funding	• Contribution by CSIR
	• Contribution by UP Government
	• Contribution from other bodies
	• Contribution from international sources
	• Contribution from bilateral sources

(Cont...)

(Contd. Table 3)

Main Task	*Details*
Civil Engineering Construction Work	• Hostel-cum-Workshop Buildings (First Phase) • Workshop Building (Second Phase) • Additional Hostel Construction (Second Phase) • Assembly and Test Laboratory (Second Phase)

Precision Engineering and Fabrication Facilities	• Survey of existing facilities in India
	• Report on existing facilities in India
	• Designing of the Nursery Workshop
	• Procuring of Machinery and Equipment
	    • Gift or Loan
	    • Purchase

Planning of residential infrastructure	• Kitchen facilities
	• Recreational facilities, sports, etc.
	• Hostel facilities
	• Guest room facilities
	• Framing of rules, procedures and practices for maintenance of
	    • services and administration
	    • transport and communications

Planning of Workshop Management	• Installation of machinery
	• Maintenance of machinery
	• Determination of the core-strength of staff
	• Determination of limits of floating staff
	• Formulation of rules, procedures and practices for
	    • maintenance of workshop services
	    • management

(Cont...)

Main Task	Details
Planning of General Engineering Services	• Electric service station • Water and sewage services • Maintenance and development of estate • Formulation of rules, procedures and practices for • engineering services
Manpower planning and mobilization	• Mobilization of contract manpower • Mobilization of floating manpower • Recruitment of infrastructural staff • Recruitment of apprentices
Procurement of machinery, materials and equipment	• Purchase of stores and machinery for workshop • Purchase of stores and machinery for services • Building of buffer stock of materials • Advance planning of future purchases
Building of the skill base	• Development of training programme • Compilation of training software • Hindi translation of the software • Creation of high precision facilities
Building of the technological base	• Creation of glass working capabilities • Development of rationalization capabilities • Development of standardisation capabilities • Creation of metrological capabilities • Planning of continuing training and summer workshops

(cont...)

Main Task	*Details*
System planning and management	• Teaming of skills and techniques • Rationalization of material management • Rationalization of production economy • Experimentation with alternatives
Optical systems	• Research in optical systems • Optical systems design and standardisation. • Rationalization of elements and components • Rationalization of tools • Rationalization of glass types
Building of information base	• Choice of optical systems for production • Acquisition of scientific information • Acquisition of technological information
The first optical factory	• Acquisition of market intelligence • Training of apprentices • Selection of factory site • Development of factory estate • Building of hostel-cum-workshop • Preparation of the project report • Preparation of the feasibility report • Financial mobilization • Start of the factory

Brief Objective

Establishing an optical component industry in UP hills making use of the ordnance capability at Dehradun in accordance with the path of technology transfer as outlined in Table 4. The next table (Table 5) illustrates the locations of the optical factories which the TNORD will set up in course of time. The next table describes the organisational and operational details already finalised. These deep planned organisational and operational measures will help self-organisation of entrepreneurships which do not need deep planning.

Table 7.4
Technology Transfer Path: Using Defence Technology for Building Civilian Industry Through Live Perspective (Vision) Research

DEFENCE STEP				CSIR STEP				COMPLETION STEP		
		PRE-TNORD INNOVATION↓						POST-TNORD ACTION		
OFDD	⇒	IRDE ⇓	⇒	CSIO ⇓	⇒⇒	NPL ⇒	⇒⇒	TNORD ⇒⇑	⇒	INDUSTRY
		⇒	⇒	⇒	⇒	⇒	⇒	⇒⇑		
		↑ POST TNORD INNOVATION								
1942		1955		1961		1966		1991		

Table 7.5
Setting up of a Technological Nursery to Mother 7 Factories as Illustrated in the Following Figure

Garhwal Region ⇓		Kumaon Region ⇓	
UTTAR KASHI ●	CHAMOLI ●	PITHORAGARH ●	
TEHRI ●	PAURI ● SITE AT SANGLIYA ⊗	ALMORA ●	
DEHRA DUN □		□ RAM NAGAR	NAINITAL ●

⊗Location of Lens Nirman Prashikshan evam Utpadan Sansthan.

- Tentative locations of optical factories to be generated from factory prototypes generated by the above Sansthan.
- Support institutions (IRDE, Dehradun and ESTC Ram Nagar).

Table 7.6
Operational Measures already Finalised

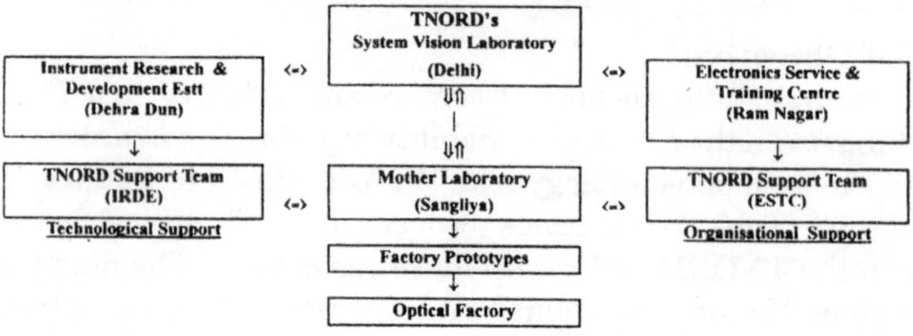

The seminar-cum-workshop on technological nursery was a very significant event. It established the commitment of the state government of UP and the scientific establishment of the country—in line with the 1958 Scientific Policy Resolution. We may conclude this reference to TNORD by reproducing following text of the letter from Shri N.D. Tiwari, M.P., to Shri I. K. Gujral, Prime Minister of India. (L.N.-3713/M.P./97 Dated 8.5.97):

"During my last tenure as the Chief Minister of Uttar Pradesh, I was requested by the National Physical Laboratory, New Delhi to inaugurate a week-long Seminar-cum-Workshop on Technological Nursery for Optics Research and Development. I was told that Professor K. J. Rosenbruch of PTB Germany, who was the chairman of the Optical Instruments Committee of the International Standards Organisation, had been specially invited to help in the planning and implementation of the programme. It was an event in which, along with the celebrated international authority in optics, were delegates from UP hills and the Gram Pradhan of village Sangliya who had donated land for TNORD. I accepted the proposal without any hesitation and I took with me the senior officials of the concerned State departments.

The event was being cosponsored by all concerned authorities not only in the field of optics but also in the concerned socio-economic and business areas. It was decided to get TNORD registered as an independent body on the pattern of CSIR functionally and INSA organisationally. NPL did all work related to the drafting and approval of the Memorandum of Association and the registration of TNORD as a Society. On its formation the NPL activities on design and development of optical systems were transferred live to TNORD.

TNORD suffered a bit of setback due to political developments in the country. But it continued to function even under this grim situation thanks to an extraordinary contribution of its system director and the support given to him by his family, friends and the Members of GB. During this period the TNORD maintained a rate of 1.5 outgoing letters per day and a research paper a month. This enabled the organisation to interact with system as a recognised SIRO.

The project is being fully supported by all the Members of Parliament from Uttarakhand irrespective of their political linkages. I would request you to take appropriate interest in the further development of TNORD."

Who did What and Who will Do What?

The above commitment of the UP government for this science driven programme and the support of all the Members of parliament from UP hills offered to the new organisation a windfall opportunity which could have been utilised by the science departments. The implementation of the TNORD programme was a simple thing in the light of this support. Some of the MPs and former MPs like Harish Rawat, B. C. Khanduri and Manohar Kant Dhyani provided strong support to the programme and the support was echoed from many other quarters. But the implementation process experienced a very tough resistance. Who resisted the programme and why? It is to find an answer to this question that we discussed earlier in this Chapter (i) confidence as resource for change, (ii) elementary industrial laboratories, (iii) structural resistance as a resource, (iv) holism of development, (v) testing the Concept, and (vi) Focus on TNORD before coming to TNORD as a Deep Planned Enterprise.

Table 3 gives details of the main tasks is details. 'Who did what and who will do what' had also been formulated at the start of the programme but not included in the table because of the flexibility of finding the most appropriate agents to implement the individual details. NPL had agreed to transfer its activities on design and development of optical systems. This simple process, suggesting a minor innovation, was a role model for establishing relationship between organic and mechanical form of organizations within the structure of TNORD. The top part of the agenda of the programme runs as follows:

But the things did not go as planned. The funds did not come as planned and the organisation had to limp on a tight rope. This limped walk on tight rope is an extraordinary achievement—it gives the programme an extra—ordinary status which perhaps no other organisation has.

Table 7.7
Who Did What and Who will Do What in the Top Part of the Tasks of Table 3

Main Task	*Details*	*Who did what and who will do what*
Organisation	• Broad acceptance of Detailed Project Report	NPL
	• Creation of Joint Implementation Machinery for the project:	NPL
	• Steering Committee	NPL
	• Memorandum of Association	NPL
	• Constitution of the Joint Body	NPL
	• Registration of the Joint Body	NPL
Funding	• Contribution by CSIR	System Vision Laboratory, TNORD
	• Contribution by UP Government	System Vision Laboratory, TNORD
	• Contribution from other bodies	System Vision Laboratory, TNORD
	• Contribution from International Sources	System Vision Laboratory, TNORD
	• Contribution from bilateral sources	System Vision Laboratory, TNORD
Civil Engineering	• Hostel-cum-Workshop Buildings (First Phase)	TNORD
Construction Work	• Workshop Building (Second Phase)	TNORD
	• Additional Hostel Construction (Second Phase)	TNORD
	• Assembly and Test Laboratory (Second Phase)	TNORD

(cont...)

Main Task	*Details*	*Who did what and who will do what*
Precision	• Survey of existing facilities in India	TIFAC-CSIO Vision 2020 Report
Engineering and Fabrication Facilities	• Report on existing facilities in India	TIFAC-CSIO Vision 2020 Report
	• Designing of the Nursery Workshop	Project Report/IRDE
	• Procuring of Machinery and Equipment	Project Report/IRDE
	• Gift or Loan	TNORD
	• Purchase	

Discovering Hidden RD Tasks

It will be observed from the above table that the first task that of building the organisation was accomplished by NPL. This was the positive part of the organisational aspect of social system. There was an under current of forces that opposed the formation of TNORD. As soon as the shot was fired, this under current of opposing forces took the command. Taking the benefit of the arrival of new director, NPL itself worked as an under current against TNORD. This situation added a new step to the above table in between the first two main tasks which may be defined as follows:

Table 7.8
Additional Task to be Added to Overcome Formidable Structural Resistance

Recoil reaction	• Contribution by CSIR	System Vision Laboratory, TNORD
	• Contribution by UP Government	System Vision Laboratory, TNORD
	• Contribution from other bodies	System Vision Laboratory, TNORD
	• Contribution from international sources	System Vision Laboratory, TNORD
	• Contribution from bilateral sources	System Vision Laboratory, TNORD

This struggle against the formidable structural resistance provided TNORD with an extraordinary opportunity to do research live in a new area. The experience was reported successive annual reports of TNORD the issues researched (Table 9).

Table 7.9
Focus of the Live Research Against the Formidable Structural Resistance

Report No.	*Year*	*Focus Title*
1	1990-1991	Background
2	1991-1992	NPL-TNORD separation: Transfer of relevant NPL activities to TNORD
3	1992-1993	Acquisition of Independent SIRO status
4	1993-1994	Focus on organizational RD and knock research in development
5	1994-1995	Casting TNORD programme in accordance with the DST format
6	1995-1996	Change of guards problem and sustenance of organisational coherence
7	1996-1997	Double-loop RD learning: Width and Depth of the TNORD programme
8	1997-1998	Organisational basics of Developmental Innovation
9	1998-1999	TNORD as a comprehensive treatise in experimental development
10	1999-2000	From NPL to NLP: Confidence inspires Confidence
11	2000-2001	Working for mindset change in science
12	2001-2002	Organisational techbio concepts
13	2002-2003	The Business of i-business

The phenomenon represented in these ten reports is most significant. How did TNORD survive this long to make the contribution? TNORD was forced by the given organisational system of science to depend entirely on the System Director who had to use his personal resources for the survival mission. He had to spare more than a half of his flat for the registered office of TNORD and invest his retirement benefits to run the office of TNORD efficiently. This placed his family members under a very unusual stress. Without this arrangement TNORD would have met the infant death. This has become a very important development and it raises many questions related to science driven development. Several new terms and new concepts have come to being to define the features of the developmental phenomenon through which the country is passing Archimedean le Lever for development.

The unusual existence gives TNORD a very special organisational position which is not shared with any other organisation-governmental or non-governmental. The role of TNORD during this period may be termed as Archimedean Lever for development because it worked from outside the gravity of science organisations. It was driven to this role by NPL initially, by CSIR and DST subsequently. Evasion of responsibility, passing of responsibility to others, and blaming everybody else for the failure are common features of the social environment in which we live. Individuals in position try to enjoy power, money and status, claim wonders they achieved and blame the rest of the society for all ills that we see around us. The System Director of TNORD not only invested his retirement benefits to run the office of TNORD but also worked hard, present a confident face, won confidence of his family members to begin with, of the members of GB, of the Members of Parliament from UP hills, of the people in and around Sangliya—the site of TNORD in UP hills, of the organisations working for the development of UP hills, and so on.

The availability of token grants from the DST and UP government had to placed certain conditions on TNORD but it was not difficult to accommodate them. This suggests that it is possible for the organic form of organisation to accommodate mechanical form of conditions. The reverse can also be made true by slackening controls at the functional end of an organisational system working in mechanical mode.

Where and Why Science has to Intervene?

The discovery of Archimedean Lever by TNORD's system vision laboratory (Delhi) is not only useful in the particular affairs of TNORD but also in the total context of rural industrialisation. The structural resistance to this science driven case example of rural industrialisation from the science organisation rather than socio-economical, political or bureaucratic circles. This is a very startling finding. This suggests that if rural industrialisation is to be made a reality the entrepreneurship has to be developed in the scientific departments who drive funds to the process. Rural development is a vision comprising missions and individual projects. The projects form part of missions and missions of the overall vision. This hierarchy is not generally understood. There may be many successful projects. The success of these individual projects does not guarantee completion of missions nor do the successful missions amount to fulfilment of the vision.

In his Zaheer Memorial Lecture (1994) Dr A. P. J. Abdul Kalam observed: "Technology development in India has progressed in many fields. In the past three decades that I have worked in Indian Space Research Organisation and Defence Research and Development Organisation, I had seen some individual visions enthusing the lives of large number of bright professionals and creating unprecedented things in the country. A vision normally generates multi-missions and hundreds of projects..." What happens when the funding agency is not able to distinguish between visions, missions and projects and treat visions and missions as ordinary projects? This suggests that the success of rural development depends on the role science plays therein and this role of science is determined in the corridors of the Department of Science and Technology — the nodal agency for science driven development of the country.

There are not many societal missions in the country. This societal missions may be regarded as simple projects that work as role examples of what can be done. Dr. Kalam admits this in his Introduction of this lecture by (81st Indian Science Congress on 4th January 1994 at University of Rajasthan, Jaipur).

> "Now and then, an important question is being asked: our Satellites are providing meteorological data, making communication link with entire world, beaming TV programmes; our Atomic Energy Research and Development Programme is providing not only certain level of power to the industry but

also radio isotopes for Radiation treatment; the Defence Research and Development has made our own tanks, missiles, and the light combat aircraft is about to come. There is a phenomenal growth of high technology behind these programmes. Did this technological growth make any impact on the society? My presentation today takes this aspect of Research and Development and deals with how a multi-organisational cooperation can transform the technology into societal missions. I am delighted that I am speaking on this topic in the city, where Dr. P. C. Sethi created history by his 'Jaipur Foot' and pioneered the concept of societal missions in the country."

The Jaipur Foot—pioneering concept of societal missions in the country—reduces to a simple project because of its isolation from the overall vision of the country's development. The mission is to help the handicapped—particularly those who lose their limbs in war or accidents. There is a reasonable support for such welfare oriented activities from the society and the government. There is no opposition to such activities. What about the vision of overall development of the countryside where things are not as visible as in this case? Things are opposed right from beginning for belonging to some abstract idealism or craze. It is here that science has to intervene. Abstraction and vision are understood only in the language of science as only those who understand this language know the route which is not visible to the ordinary beings.

Rural Development as Basic Research

Fortunately for us this requirement has been appreciated in the Scientific Policy Resolution and in the concept of scientific temper—the issues which have been discussed earlier in detail in this chapter. The point has further been elaborated in the deep planned TNORD programme.

During the course of the struggle of TNORD it was found that the research investigations related to rural development closely follow the pattern of basic research rather than applied. It was also found that the results of organizational research of the corporate world are equally applicable to rural development because both these sides are vitally interested in change management. The interest of the current RD activities lies in the status quo and it resists change. It considers research topics in their fragmented subject form, opposes holistic approaches and is implementation shy. When an activity shoots out, there is the natural recoil problem and its management is as important as the management of the activity shot. These two conclusions are not reported in ordinary R&D literature related to rural development. As we have discussed earlier in this note the corporate literature has discussed structural resistance at length but this discussion is limited to the mechanistic managerial language. The literature awaits inputs from developing countries to develop fuller insight.

The importance of this recoil reaction is not ordinarily appreciated—it comes up during the implementation stages and not during the formulation stages. Pessimistic colonial culture still persists in which nothing was implemented except law and order and the hidden agenda of the colonisers. It seems that science in India is being managed in the court culture of the past few centuries and the current trends are inconsistent with the basic roots of Indian culture. The political circles and enlightened activists are aware of this cultural limitation of the process of development and they often go the other extreme and there by overshoot the target. It seems that scientists up in hierarchy are interested more in managing the fund rather than using it for societal purposes. The process of fund management uses usual techniques of committee rule and procedures to provide funds to some and deprive the others forgetting that funds are provided for social function and not for performing rituals in fund allocation. Are the authorities aware of this fundamental weakness? Is somebody looking for new answers to old problems?

Looking for New Answers to Old Problems

The ground reality that efforts for rural development made so far have not succeeded has been accepted at the highest governmental levels. *Times of India* news report (dated 19 December 1999) on the 79th Annual General Meeting of ASSOCHAM quoted observations of the Prime Minister in following words: "The PM made it clear that even in a knowledge-based society, the objectives would remain unchanged: poverty eradication and an equitable social order. The knowledge base would help find "new answers to old problems", faster and more effectively". As the older measures have not succeeded every new measure deserves serious consideration and in this light the above observations of the PM are important. The members of the corporate world are working for survival through intense R&D in a positive assertive sense. Change is the focus of their activities. The PM seems to have taken a measure for poverty eradication and creation of equitable social order by appointing the key scientist Dr. Kalam as principal scientific adviser and grassroot worker Shri Rawat as the S&T minister. The above combination is likely to give a visionary fresh look at the entire gamut of issues related to policies and decisions concerning development of the villages.

The only solution is the knowledge empowerment of village panchayats through a network of technological nurseries in the countryside on the pattern of TNORD—the pioneer. Technological nurseries can handle these issues effectively in the following manner.

> Village-specific technologies can be identified by undertaking specific RD in this direction involving the geniuses of the soil in the technological nursery (local technology server organisation). These technological nurseries must be deep planned for some high level marketable production so that they have productive network relationship with national laboratories and research institutions to invite specialists for consultation and help in the local RD. These technological nurseries must work as learning organisations and establish self-organised culture and facilitating environment by working on appropriate role models to motivate rural artisans and entrepreneurs. For example, TNORD—the pioneering technological nursery - is planned to continually work on optics RD for upgradation of skills of optical workers and the same facilities will be directly or indirectly used to help general industrialization of the village and surrounding region. The presence of these deep planned technological nurseries designed to function in the following operational scheme which has been styled after the scheme of Chris Argyris (1994, Blackwell Business, p. 8):

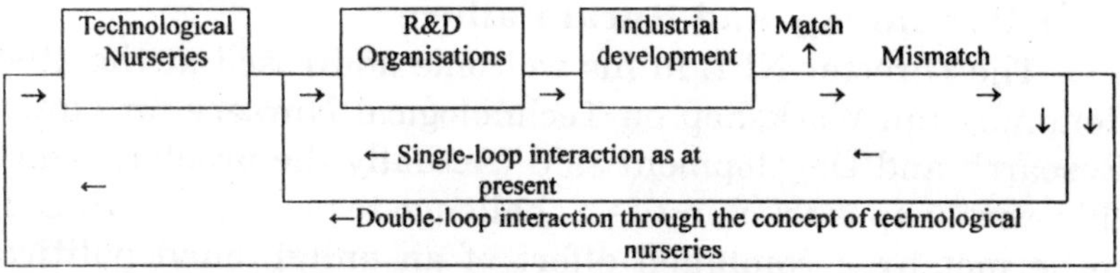

In relation to TNORD the existing R&D laboratory is the Instrument Research and Development Establishment, Dehradun and it will provide RD consultancy for the technological nursery as the first step in the process of industrial research. A rhythm of trials and errors will lead to a dead end stage. After this the technological nursery will have to come in and devise ways and means to improve things. This can be done in various ways for example by replacing R&D organisation for a while to overcome the hurdle. If the hurdle persists the technological nursery will have to create new facilities at the site to overcome the hurdle. The same process may be used to fix proper criteria for selection of other industries in the region using the same operational style. The same operational style will be used to streamline raw material availability and marketing of products. Technological nurseries are also meant to empower the village panchayats with the latest developments in the relevant fields and do adaptation R&D. The presence of technological nurseries keeps technologies moving and in step of the market needs and they are not allowed to become stagnant.

Being the pioneer, TNORD had to meet challenges as they came more with the power of mind than the power of money because of delays in receipts of grants. This experience needs to be generalised and scaled for organising training for this purpose. The training can be made relevant by technological nurseries by undertaking R&D in forward as well as backward directions. This requires continuous problem solving R&D and it can only be arranged through technological nurserie by incorporating TNORD experience in the development of its site-village. Technological nurseries have handicaps but they have strengths also and within their area of operation they should not shy away from the competitive challenge of the corporate world.

Fear Psychology and Rural Failure

The Director NPL in his welcome addressed at the 1988 Seminar-cum-Workshop on Technological Nursery for Optics Research and Development said "Actually the problem which this workshop is touching, is a challenge to planners... which is to be met by a combined effort of an enlightened political leadership and sympathetic administration on the one hand of managers, economists, environmentalists, scientists and technologists on the other and cautioned... If you have the concept left alone at NPL, I think it will wither away and may die someday, and we will be left with reminiscences..."

The follow up action on the recommendation was even harder. We are living in the age of the fittest producers of goods and services and this fitness is acquired through intense research and development in an implementation friendly environment. The old political stances of left, right and centre have been changed to new ones and the definition of science is being changed to new science.

What is necessary is to recognise the reality different between the developing and developed societies and apply the standards accordingly. Roots are roots and fruits are fruits. The organisation gets transformed with time and reaches the fruition stage. Developed countries are at fruition stage and developing countries at the root stage and these two stages are different. The experience of the developed countries can be reinvested in developing countries. Likewise the experience of the corporate world in marketing can be reinvested in rural development.

Developed markets today are working for distant marketing as the near markets are saturated. Developing markets do not suffer from this disadvantage but they have to compete with the developed markets. By following the experience of the developed markets, developing markets can turn their disadvantage into an advantage. Here comes the role of technological nurseries.

Reinvestment Strategy for Rural Development

A TNORD associate, Pallavi Dhyani, highlighted the concept of technological nurseries against the background of the book

Rural Industrialisation: A Plan for the Future (2nd revised edition 1995, Vikas Publishing House, New Delhi) and described it as a methodology to generate reinvestment skills.

The phrase 'Reinvestment skill' may be related to more familiar terms like innovation. We have noted that the phenomenon of globalisation has placed at the disposal of the researchers working on grassroot problems a vast inexpensive information on the struggle of the corporate world to survive. This embodies the experience of the entire industrial world and the researchers job is to determine how to use or reinvent it for solving the grass root problem.

Infusion of new or external knowledge is, however, not a simple thing. This has to be done through technological nurseries. Perceptions about these technological nurseries are not clear yet. There are many hidden steps between theory and practice. These are being clarified by TNORD through conscious problems solving efforts. The purpose of technological nurseries is to develop these reinvestment skills. TNORD is working to develop the reinvestment skill by attempting to produce internationally marketable optical components.

Reinvestment strategy at least solves the problem of major funding. The reinvestment strategy also needs funding to implement the outcome of the primary investment. But it saves the money and time spent on the primary projects. The corporate world is spending a lot of money to determine regular market feedback at lowest cost, necessity of quality marketing arrangements and regular communication with the consumers and traders, and on human resource development even if transport and storage bottlenecks are not serious problems for large corporations. The task of the reinvestment R&D is reduced to (1) how to make best use of resources already available in order to set new goals determined by the experience of the corporate world, (2) what new resources should be created in the given environment, (3) to create these new resources and (3) how to sustain this process of interaction on a continuing basis. The mechanics of this process has been developed in the concept of technological nurseries.

Technological nurseries may be considered as in house-R&D institutions that make existing corporations to survive in the fierce battles of the ongoing corporate war. These are known as industrial research laboratories. At the root level they become technological nurseries in the language of reinvestment strategy just outlined. The industrial research laboratories are required to work in a rapidly changing environment but technological nurseries can afford a slow pace and this is why many of the findings that cannot be implemented at the corporate level, can be implemented at the nursery level.

The Taori Singh book discusses appropriate technology (thrust area 3) in Chapter 11—Major thrust areas and overall strategy for rural industrialisation highlighting government policies and development programmes, organisational structure, role of voluntary agencies, coordination among various agencies, information-cum-knowledge centres, consultancy services, incentives, preparation of directories, training and development centres, seminars, symposia and conferences, extension work concluding that development of appropriate and simple technology indigenously would need the following changes in the scientific approach: (i) Any scientific data should be accepted as true only if it is clearly recognisable as such by practical applications in a pilot plant (The decision on scientific data should be unbiased and free from all prejudices in order to realise optimum results from such experiment), (ii) Each problem should be divided in parts and projected in a scientific manner, (iii) Scientific analysis that is understandable easily should taken up first for pilot trials and complex problems should be projected in a phased manner (At the same time, complete enumeration and comprehensive review of the techniques evolved should also be undertaken), (iv) The training institutes and equipment manufacturers should be consulted before adopting any upgraded technology.

It finally notes that the area of efficient technology is very important, and there is an urgent need to adopt a broad visionary approach in this field. What does the efficient technology imply? As the metaphor technological nursery suggested technology has to be live in order to be appropriate. It has to be continuously

efficient scientists capable of breaking through the barriers of every kind. The book thus fully supports the concept of technological nurseries in the countryside without being aware of the TNORD movement. The following diagram establishes the relationship between TNORD and the book.

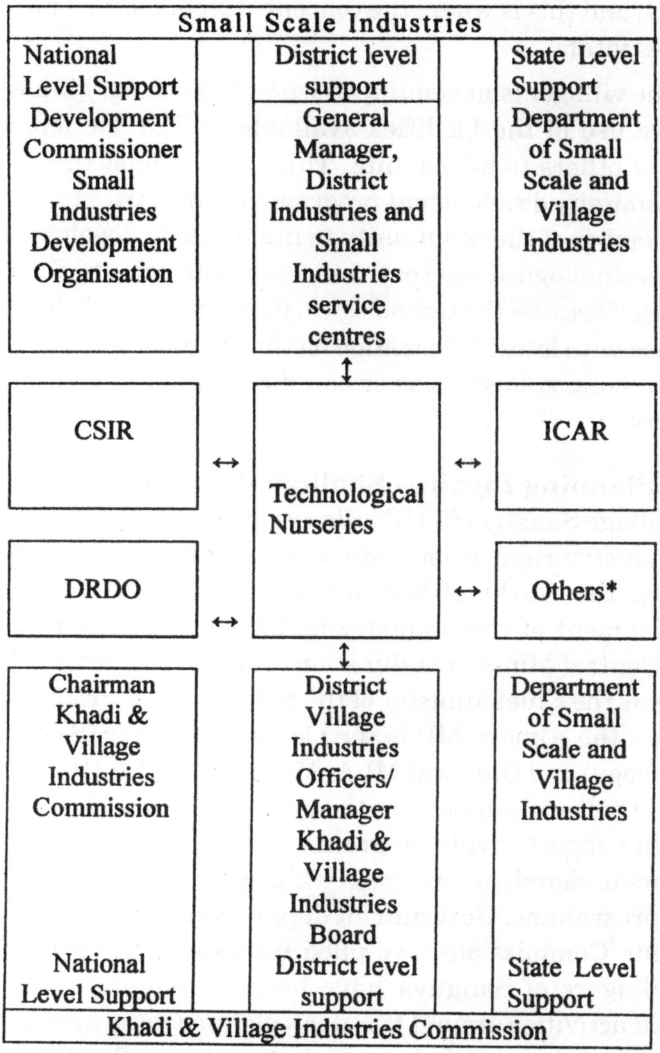

* CAPART, NSIC, SISI, NISIET, IITs, ICMR, Universities, NCERT and voluntary agencies, etc.

The above diagram discusses the current organisational structure for rural industrialisation in detail. There are existing channels of support to the villager to encourage him to take up entrepreneurship. This vertical support needs cross fertilisation in a horizontal direction by the inputs of scientific and industrial research and this is where the concept of technological nurseries becomes vital.

The villager is not equipped with the necessary knowledge to make use of the facilities available to him. He left at the mercy of others to advise him. This goes against the spirit of the community development programme which tries to empower each member of the community to find route to development. It is the technological nursery that empowers him to find route for himself because the technological nursery is a symbolic villager equipped with latest information on the progress of science. What this symbolic villager does or can do is bound to inspire other villagers.

Deep Planning Inspires Shallow Planning

Village Sangliya in UP hills are aware of the deep planned lens industry right from 1969 when there was an agreement between the Garhwal MP and Ranikhet MLA to work for establishment of this industry in UP hills. The Garhwal MP was a Central Minister (Education) and the Ranikhet (Almora) MLA was the Chief Minister of the State of Uttar Pradesh. After 30 years the Almora MP is the Central Minister (Science and Technology) and Garhwal MLA, Minister for Hill Development UP. The political support for the programme has all along been there in support of this programme. Successive generations of villagers in Sangliya have been waiting for the implementation of the programme. Government departments at centre, UP and Planning Commission have piled up files on the programme. The villagers of Sangliya have been contemplating shallow planned activities around this deep planned programme. There had been no fear psychology at the rural end. But the science departments both at the centre and state have been suffering from acute fear psychology and trying to evade it by tossing the responsibility in each others courts. One deep planned industry followed by mushrooming shallow planned industries has potential of rural transformation.

This is not a hypothetical proposition. China is creating 1 crore jobs annually almost by the identical process. All that is necessary to overcome the crisis of confidence. What is the secret of the Chinese success? It is encapsulated in the following table. The Chinese S&T expenditure is primarily invested in experimental development—the type of activities which are defined as technological nurseries in our country.

Share of national S&T expenditure by types of work in China

Type	*1986*	*1990*	*1995*
Basic research	12.4	7.3	6.7
Applied research	57.7	28.5	30.5
Experimental development	24.6	64.2	62.8
Other activities	5.3	-	-

It may not be possible to realise whatever is planned. It is difficult to determine all fatal points right at the planning stage. These are determined during the implementation process. This is why successes as well as failures result from the planned programmes. It is important to overcome the fear of failure. This simple understanding makes all the difference. West succeed because they took the risks. Japan succeeded in the same way. China is doing the same thing now.

Conclusion

Any organised effort to develop the rural industries cannot ignore the above role of technological nurseries or of experimental development in Chinese S&T vocabulary. Without organised RD effort in this direction the villagers seem to have been left with no other option but to shift to alternative vocations or to migrate to urban centres to earn their livelihood and the trend is tending to deny the villages a major element of self-sufficiency which they had enjoyed for ages. The above and other related considerations call for a fresh look at the entire gamut of issues related to S&T policies and decisions concerning development of the villages. These policies must be modified to ensure reinvestment of existing skills by developing reinvestment skills through technological nurseries or experimental development. Skills must not be allowed to become redundant— they are to be reinvested through continuous reinvestment experimentation. 'Through experimentation we evolve; through stagnation we die', says Roger Bennett, Series Editor, The McGraw-Hill Training Series of the McGraw-Hill Book Company Europe.

The concept of technological nurseries for rural industrialisation and entrepreneurship development, which is being nurtured through the strategic deep planned pioneering Technological Nursery for Optics Research and Development (TNORD), is an extension of the Scientific Policy Resolution which the Indian Parliament passed in 1958. The concept uses the power of human mind for the effective combination of three factors, technology, raw materials and capital. The TNORD example demonstrates how creation and adoption of new scientific techniques can make up for a deficiency in natural resources, and reduce the demands on capital. The country has a huge infrastructure for scientific research and development and it supports the world's third largest scientific manpower. This infrastructure can be mobilised for the reconstruction of the country's neglected countryside — the *Maila Anchal* (dirty region) according the Novelist Phanishwar Nath Renu through a deep planned network of technological nurseries. This can be a powerful strategy for the perspective defined by the Scientific Policy Resolution.

Appendix A

Budget Estimates (Initial Five Years) as Given in Detailed Project Report of TNORD

(Rupees in lakhs)

Head of Expenditure	*Year*					
	1st	*2nd*	*3rd*	*4th*	*5th*	*Total*
Salaries, wages and Honoraria	3.00	6.00	9.00	11.00	13.00	42
Works	14.00	12.00	12.00	6.00	6.00	50
Services	2.00	4.00	4.00	1.00	1.00	12
Equipment and Machinery	16.00	16.00	8.00	4.00	4.00	48
Chemicals and Apparatus	4.00	4.00	4.00	2.00	2.00	16
Contingencies	0.8	0.8	0.8	0.8	0.8	4
Maintenance	0	0	0.4	0.8	0.8	2
Miscellaneous	2.00	2.00	2.00	2.00	2.00	10
Total	**41.8**	**44.8**	**40.2**	**27.6**	**29.6**	**184**

Notes

1. The expenditure will be shared on 50:50 basis by DST and Hill Development Department/UP Government.
2. The main investment on the part of the State Government will be on civil construction work at the site.
3. The main investment on the part of DST will be on machinery and equipment and the overall infrastructure.
4. Expenditure on additional facilities such as P.K. Kichlu Museum, Library and Hobby Centre is not indicated in the above table.
5. Expenditure on optical factories has also not been included in the above table.
6. The expenditure incurred by supporting organisations, namely, Instruments Research and Development Establishment, Dehradun and Electronics Service and Training Centre, Ram Nagar is not directly included in this table.

Appendix B

Schematic Flow Chart of Technological Nursery for Optics Research and Development (TNORD)

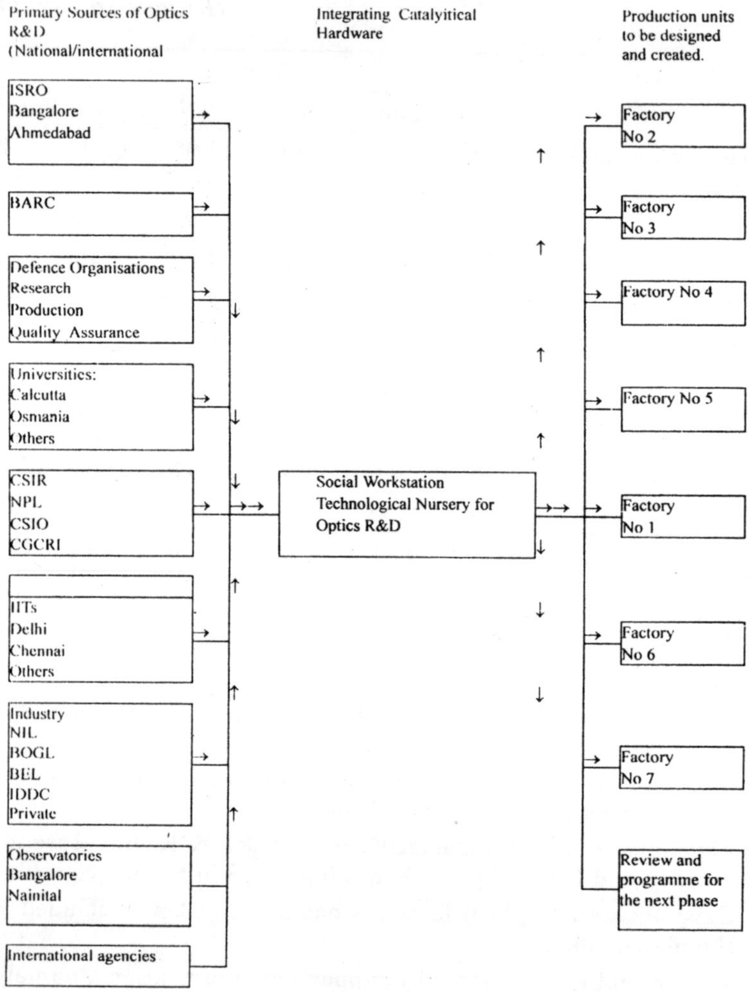

8

New Technology Options for Rural Industrialisation

A.K. Bhatia

Thanks to the Green, White, Yellow and Blue revolutions that led to significant improvements in agriculture and allied sectors in the country. However, the rural non-farm sector has not benefited much by these revolutions; the poorest of the rural poor continue to be in this sector. This calls for promotion of cottage industrial enterprises in rural areas to provide the much needed avenues for employment and also help in value addition to the local resources. Thus, a radical approach is needed to develop innovative technologies and to transfer, transform and translate these.

Over the years, the national laboratories under Council of Scientific and Industrial Research have developed a large basket of technologies, many of which have special significance for the rural sector, e.g., value addition, income enhancement, employment generation, drudgery mitigation and improvement in quality of life through low cost shelters, potable water, health, nutrition, etc.

Salient rural technologies are briefly described. These are for processing of raw material near the source, potential market of finished products and upgradation of local skills leading to new vistas for rural industrialisation.

Introduction

Three-fourth population of India lives in rural areas. A majority of them, nearly 80 per cent, are engaged in agriculture and allied activities. As a result of growth of economy, the share of agricultural employment is declining whereas the rural work force is growing at 1.8 per cent per annum Additionally the marginal farmers also need supplemental income. The strengths of this segment of population are:

- a large and growing reservoir of potential workers,
- abundant resources of raw materials (agricultural produce, agrowastes, forest produce, livestock, mined materials, etc.),
- traditional skills and capacity for initiation of viable small enterprises, and
- large and increasing demand for goods and services.

Above scenario calls for increased efforts on economic development of rural areas through S&T inputs so that the material resources available there can be processed in small units right there, for consumption locally as well as otherwise. This would provide much needed avenues for employment generation, especially in the non-farm sector and help in value addition to the endogenous resources. It is generally observed that the traditional cottage industry which is mostly based on skills acquired outside the formal school system, suffers from technological obsolescence, low productivity, inconsistent quality, drudgery and dependence on weather conditions. It is in this context that a radical approach is called for to develop innovative technologies and to transfer, transform and translate these with adequate safeguards against technological unemployment, environmental pollution and exodus from traditional homesteads. This would open up new vistas of rural industrialisation. The salient ones are briefly described, as under:

Food Processing Technologies

Today, India is the largest producer of milk in the world. It has also emerged as the second largest producer of fruits as well as vegetables with annual production of 45 and 90 million tonnes respectively. The country has entered into an era of surplus in foodgrains with production crossing 205 million tonnes. With the array of natural products, bestowed through varied agro-climatic conditions, the agro-processing sector has to play a premier role. A few important technologies for small/cottage sector are:

1. **Mini Rice Mill:** A mini-rice mill has been developed which saves on valuable rice bran that provides 14-18 per cent edible oil. In addition, brokens are reduced to 0.5 per cent and milling time reduced by 25 per cent.

2. **Mini Pulse Mill:** The hand operated pulse dehusker can process 60 kg pulse per hour. A motorised version, working on 1/2 HP motor, dehusks 150 kg pulse in an hour and minimize the brokens. A composite mini-pulse mill is also available that facilitates all the operations in a single unit including dehusking, aspiration and separation of different components.

3. **Oil Expeller-cum-Paddy Dehusker**: This simple machine, named 'Grihashree', overcomes day-to-day household drudgery of rural women. It is used for dehusking of paddy as well as for extraction of oil from the oilseeds at domestic level. The capacity is 18 kg rice/hr or 900 ml oil/hr by motorised unit. The manually operated unit yields 3 kg rice/hr or 125 ml oil/hr.

4. **Mini Wheat Mill**: It provides atta, suji, maida and bran simultaneously. So far, suji and maida were manufactured in large mills requiring high capital and sophisticated technology.

5. **Parboiling of Paddy**: The improved technology overcomes the problems of prolonged soaking and foul smell. It provides higher yield of head rice and better storage characteristics.

6. **Quick Aging of Paddy**: The rice from freshly harvested paddy cannot be consumed immediately and needs to be stored over a few months to improve cooking characteristics. The similar quality can be attained now in a few weeks time with simple processing.

7. **Bakery Products:** Despite the advent of semi-automatic and automatic bread lines as well as biscuit plants, a sizeable cross section of the population still prefer fresh cottage bread. A low cost bakery unit with improved technology can be set up in semi-urban as well as rural areas.

8. **Instant Pickles**: A new concept has been worked out for making an instant pickle mix which can easily be reconstituted into a ready-to-serve pickle by the simple addition of oil and water. The dry mix has the features of convenience, better keeping quality, reduced packaging and transportation cost.

9. **Evaporative Cool Stores**: Evaporative cooling (EC) system does not depend on conventional energy sources. The shelf life of commodities is extended by 20 per cent.

10. **Tutti-fruity**: The raw papaya is largely used to make tutti-fruity required in bakery, sweetmeats, ice creams, salads and paan. The demand of this product is rapidly increasing. A number of small scale units are coming up to manufacture these items. Tutti-fruity has a lot of demand.

11. **Low Cost Preservation of Eggs:** In order to enhance the shelf life of eggs and to tap far hung markets, a simple process is now available. The eggs are coated with a oil formulation whereby the shelf life is increased from the normal 5-6 days during summers to well over 10-12 days at 38°C, 4 weeks at 25°C-30°C, 12 weeks at 13°C and 24 weeks at 7°C. There is a wide scope for a small units providing this service at nominal cost.

12. **Poultry Dressing**: Due to the rapid growth in broiler production and consumer requirements, the hygienic dressings, packing and distribution of poultry meat have become a necessity. Further, it has the advantage; of utilizing the waste.

13. **Rabbit Farming**: Rabbit culture provides valuable meat, and also the skins for fur industry and the wool for the hosiery industry. One does produces 30-40 more rabbits during a year. Important fur and meat varieties are New

Zealand White, Giant, Soviet chinchilla whereas Angora is a wool variety, used to manufacture prized shawls, blankets and woollen garments.

14. **Quail Farming**: Commonly known as '*Bater*', its meat is highly cherished as a table delicacy, Farming of Japanese quail has opened up new vistas in poultry production and as a remunerative rural activity. The quails are good source of egg, meat and for sport.

15. **Animal Feed:** Cattle and poultry feeds are the mixtures of various ingredients like maize, rice bran, oilcakes, molasses, fish meal, etc. The composition can be varied depending on the local availability of raw materials.

16. **Leaf Cups and Plates**: Leaf cups and plates are traditionally made in Indian villages by hand. The laborious craft of making leaf cups and plates is now transformed through a machine operation. The machine can make these articles in various shapes with better quality and dimensional stability. An operator can produce about 2,500 cups/day from thin leaves and about 1,000 from thicker material.

Speciality Crops

There is an ever increasing demand of natural plant materials for use in the manufacture of drugs, pharmaceuticals, cosmetics, food flavours, perfumes, soaps, paper/pulp, other industrial products and as fodder, fuel, timber etc. Cultivation and processing of these plants has considerable potential as a small scale industry. In addition, wastelands can also be brought under cultivation. Depending upon landholding, processing can be taken up on individual or cooperative basis and in the industrial sector. Some of the important plants are listed below:

Medicinal Plants

- **Dioscorea composita and D. floribunda**: Raw materials for diosgenin for further synthesis of steroidal drugs.
- **Solanum khasianum**: A source of solasodine for production of steroidal drugs.
- **Sarpagandha (R. serpentina)**: The roots are used in the manufacture of antihypertensive and sedative drugs.

- **Aswagandha (Withania somnifera)**: For treatment of rheumatism, gout, hypertension.

Aromatic Plants

- **Lemongrass (C. Flexuosus)**: Used in flavouring, cosmetics, soap, perfumery, synthesis of citral needed in the manufacture of Vitamin A and a number of aromatic compounds.
- **Java Citronella (C, winterianus)**: A source of perfumery chemicals like citronelial and geraniol, needed in soap, perfumery and flavouring industry.
- **Palmarosa (C. martinii)**: A rich source of high grade geraniol.
- **Peppermint (Mentha piperita)**: A source of peppermint oil needed in flavouring and pharmaceutical industry.
- **Patchouli:** The leaves yield a valuable oil that is used practically in most of the perfumes as base because of its fixative property.
- **Geranium (P. graveolens)**: A source of geranium oil needed in soap, cosmetic and perfumery industry.

Other Economic Plants

- **Fast Growing Sps**: The fast growing plants have assumed increasing importance due to the depletion of green cover. These provide fodder, fuel, timber, biomass for paper, pulp and other valuable industrial products. These can also be adopted in various wastelands, e.g., alkaline, saline, water logged areas, marshes, swamps, jhum lands, sandy, clayey soils, gullies, ravines, denuded and eroded soils, mined wastelands, etc. The plants possess fast rate of growth, need less inputs and improve soil and environment.

Leather and Animal Waste Based Technologies

Leather processing is an another important area which is largely in the hands of unorganised sector and technology upgradation can make a marked impact on value addition.

1. **Tanning of Hides and Skins**: Conventionally the rural tanners use the pit or bag method for tanning which is drudgerous, time consuming and results in lower quality. A simple manually operated drum system for rural tanning is now available for tanning of hides and skins with a capacity to tan 2 hides or 20 skins in a batch. This technique has been well accepted in rural tanning operations.
2. **Wet Blue Chrome Processing:** It is a technology that brings in marked improvement in the traditional cottage tanning operations. It results in upgraded product quality as well as saves on the processing time. The raw hides thus tanned can be further processed into chrome tanned leather.
3. **Milled Grain Side Leather**: It is a buffalo hide based leather used for making travel goods, hand bags and shoe uppers. Usual tannery equipment is used.
4. **Hydrolyzed Feather Meal**: Poultry feathers create problems during broiler dressing and are not digestible in raw stage due to disulphide linkages of protein molecule. These are hydrolyzed under optimum conditions to obtain protein rich meal for the feather.
5. **Improved Machines and Tools for Leather Artisans**: The improved tools help in reduction of drudgery, improvement in productivity and quality as well as additional value earnings for the rural leather workers. The improved tools have been well accepted by the artisans.

Low Cost Building Materials, Components and Systems:

Shelter is one of the basic needs of a human being. The construction materials and components have a vast market which has further risen in recent years with emphasis laid on shelters for economically weaker sections under various welfare schemes. The rural artisans can play a vital role in taking up activities related to their manufacture, marketing and laying new systems.

1. **Brick Moulding**: A simple hand moulding table has been developed that moulds bricks in accurate shape and size. This device can mould 1,000-1,500 bricks in a day. Improved dimensional accuracy saves in mortar consumption.

2. **Stabilized Mud Blocks**: Soil stabilization implies improvement in strength and resistance to erosion. A coating of bitumen/asphalt over the soil particles provides additional cohesive strength and reduces water penetration in the mass. Stabilization achieved through the addition of 4-10 per cent cement, greatly improves the water resistance and strengthens the soil mass.
3. **Precast Stone Blocks:** Precast stone block masonry is an economical substitute for random rubble masonry. The blocks of standard size (30 × 20 ×15 cm) are made of cement concrete mix and stone pieces upto 25 cm size.
4. **Hollow Concrete Blocks:** It is a low cost item with adequate strength for building construction. It eliminates plastering as well as rendering and reduces dead load of the structure.
5. **Bricks** from inferior soils, alkaline soils, acidic soils, alluvial soils, etc.
6. **Clay Roofing Tiles:** Mangalore-pattem clay roofing tiles are quite popular in the country. Production of such tiles from alluvial soil could not be undertaken earlier because of the poor workability and silty nature. A process is now available for the manufacture of these tiles from alluvial, red and black soils in small scale sector.
7. **Clay Flooring Tiles:** The improved process yields tiles with uniform texture and colour, a metallic sound and good finish.
8. **Medium Density Fiber Boards:** Various types of agro and agro-industrial wastes are converted into fibers of required size. The product is waterproof, termite resistant and fire resistant.
9. **Woodwool Boards :** The woodwool is saturated with cement slurry and compressed in the form of a board. Due to its medium density (300 to 500 kg/m^3) and open texture, this board is primarily used as a thermal and acoustic insulation material. Other characteristics are its inherent resistance to fire and termite attack.

10. **Frameless Door and Window Fixtures:** Frameless doors and windows effect good saving in timber and overall cost. Special type of hinges are used for fixing shutters without frames.

11. **Solar Timber Seasoning Kiln:** Before use, fresh timber is required to be seasoned by storing in a shed for prolonged periods. This timber seasoning can be accomplished in half the time in a solar kiln due to higher temperature inside the glass chamber.

12. **Non-Erodible Mud Plaster:** The mud plaster over mud walls gets eroded during rains and necessitates costly repairs. A simple plaster is applied on mud wall surface which enhances the durability of mud wall and reduces the maintenance cost.

13. **Plinth Protection of Mud Walls:** The lower portion of mud wall gets eroded during rains due to the direct striking of splashing water or water flowing in streets. This causes sudden collapse of the house. A new method provides protection of such walls to withstand rain and the flood. The roof is finished with non-erodible mud plaster.

14. **Fire Retardant Thatch Roofing**: The thatch roof made by the traditional method is generally loose and deteriorates fast. Spread of fire in such roofs is also very fast. In the new method, the roof is finished with a plaster to make it fire retardant and water repellant.

Drinking Water

A large number of villages still do not get adequate supply of potable water. A sizable requirement is met from ground water sources. On account of indiscriminate utilization of this important resource, both the quality and supply position has been deteriorating. Technologies are now available for exploration, purification, quality assessment and storage of water.

1. **Groundwater Exploration**: Wells have been used as source of water supply since time immemorial. Locating the exact spot for exploration of groundwater has remained a tricky business. Today simple technology is available where ground water exploration is possible with 95 per cent success ratio.

2. **Rainwater Harvesting**: Rainwater harvesting is one of the most economic systems for collecting drinking water in areas getting sufficient rainfall. The technology has been modified for gathering water run-off based on micro-catchment principle. It has been well adopted in Meghalaya, Manipur and Tripura.
3. **Water Filter Candles**: Water filter candles, manufactured by potters are fitted to domestic water containers, including earthen pitchers.
4. **Amrit Kumbh**: It is a simple, efficient and durable filtering device for the complete removal of bacteria and reduction of flouride, sulphate, lead and organochlorine pesticides from water sources.
5. **Handpump Attachable Iron Removal Plant**: Large population in the country depends on groundwater having characteristic high iron content. A simple plant has been deviced to remove excess iron to serve a population of 250.
6. **Defluoridation of Drinking Water**: Excess fluoride in drinking water (above 1.5 mg/L) causes both dental and skeletal fluorosis. The domestic treatment is performed in a bucket or any suitable container.
7. **Water Quality Testing Kits**: These are low cost portable kits to test the quality of drinking water for routine; chemical and bacteriological analyses.

Environment and Sanitation

The need for low cost sanitation system is well-known. The artisans can play a vital role in practicing new techniques in disposal of waste water through brick in-situ system or new soak pit models, sanitation devices as well as manufacture of low cost stone sanitaryware. A number of technology options are available.

Upgradation of Artisanal Skills

The artisans have to adopt new skills to manufacture innovative items as per present day market demands that provide higher remunerations.

1. **Floral Crafts:** There is a good demand for floral handicrafts made out of dehydrated flowers and foliage. The arrangement of dehydrated flowers and foliage in sealed transparent glass or plastic jars makes very good decoration pieces with a potential market demand.
2. **Fishing Hooks:** The process of manufacturing fishing hooks employed by the artisans is based on primitive heat treatment and finishing techniques. The entire process of hook manufacture has been revamped consisting of selection of material quality, hardening, tempering and better geometrical configurations.
3. **Fur Garments from Unconventional Skins:** The new processing technology can produce value added furgoods from the skins of sheep, goat and rabbits available as slaughter house by-products. These items are produced at a cost substantially lower than conventional furgoods.
4. **Ornamental Fish Rearing and Aquarium Making**: The fascinating world of the colourful aquatic life has attracted people since age old times. Rearing of fresh water tropical ornamental fish is done in plastic or cement tanks. These are placed in glass aquarium.

Modalities for Availing the Technologies: CSIR provides the various technological details, techno-economic features, arranging demonstrations of technologies, provide training facilities, trouble-shooting and other such technical back-up.

9

Rural Industrialisation-Some New Initiatives

Dr. Rajendra Prasad

Today, the decentralised rural industry, and the so-called, modern, organised, urban industry in India are really in conflict, and competing with each other. The modern industry is displacing and destroying the rural industry. This is very unfortunate and harmful, not only to the rural industry but to the modern industry too. Both must realise their complimentary role. The modern industry can benefit a lot, if they, instead of always looking to and imitating the West, and importing the technology from there, start basing themselves on what exists in the country, and start building from there, utilising the S&T, manpower and infrastructure available in the country.

There is a need to have an indigenous base and that can be provided by the total industry, the traditional industrial system of production, which had been stronger than anywhere else in the world at one point of time, not too distant in the modern history. The need is to build upon that, improve upon that, strengthen that, remove its deficiencies and weaknesses in light of the modern developments of science and technology, in which too, the country has developed a huge infrastructure and manpower. Let the two interact properly, help each other, develop each other and thus provide the right industrial set up and base needed so much by the country today.

Target Groups, Approach and Strategy

Even by official estimate, a large percentage of our total population is still below the poverty line. Programmes like

Integrated Rural Development have, though helped to some extent in ameliorating the conditions of the poor, the problem at large remains unresolved. Undoubtedly, new initiatives at pilot programmes are critically needed to arrive at more effective solutions for poverty alleviation. Not only has there to be more improved and greater and more diversified local value addition through production and through utilisation of available natural resources for local value-addition, but also the ability to retain control over that value-addition and the resources. The rural poor covered by IRDP belong to such occupations as agricultural labour, artisans, small and marginal farmers and include freed bonded-labour, recipients of surplus land, the physically handicapped and women. It is apparent from the occupation spectrum and the scale of assistance for productive assets, that the small producers have remained small producers or petty traders while agricultural labourers are being metamorphosed into petty proprietors. Organising the unorganised workers and small producers is a key element in any genuine strategy for empowering the economically weaker. Under conditions of market competition there is no other available alternative for the small producers and the unorganised worker but to eliminate mutual competition amongst and between themselves and to associate and to co-operate to form large participating organisations which enable them to achieve greater and more effective competitive capabilities to retain and diversify their markets and to gain greater and greater access and control over available resources. In doing so the small producers must necessarily cease to be the individual, the mutually competing small producers. This is what has occurred earlier in large scale enterprises. The science based technical division-cum-synthesis of production, has generated a requirement for cooperation in production.

Apart from other aspects, the question of organising is genetically linked to technology. For the organisation of the unorganised, to be facilitated there has to be a shift from virtually isolated and scattered production techniques through transitional forms, to technology engendering cooperation in production: and the nodalising operations themselves expressed in terms of value-adding production operations/processes so that the nodalisation enriches, transforms and diversifies the existing production furthermore, the greater the degree of organisation, the greater the ability to influence the competitive market.

Along-with the organisation of the unorganised, and indeed as a part of this social organisation, there is a need for some form of multi-sectoral area planning and technological choices to enhance the local value-addition through secondary production, to effect recovery of degraded lands, allow a sagacious water management and conservation, etc. and enable appropriate linkages within the area.

The principals of cooperation provide the foundations for networking and for networking the networks to create larger areawide production Networks sharing nodal facilities/ operations/ processes to enhance the productivity of and value-addition by the constituents.

What is needed to be done is the empowerment of the people through:

(1) enhancement of the scientific and technological knowledge, and skills of the people;

(2) enhancing professional skills, par ticipative management and group entrepreneurship capabilities, and

(3) developing/adapting techno-economically and socially viable technology.

As a strategy the networked agro-industries vertically integrated with wasteland recovery, will be organising and participatively employing agricultural workers, and will provide synergising support through the consultative arrangements to the small/marginal peasantry by providing inputs for higher value-added land-usage by the latter, and by procuring their output to process through the networked production and sell the resulting products locally and non-locally. The peasants, the agricultural workers, the forest dwellers and workers, the fisherfolk, and the artisans, largely using traditional techniques for their day-to-day production and consumption needs, for their seasonal needs and for their longer duration investments such as in milch/draught cattle or on bullock-carts, etc. have 'spontaneously' formed typical patterns of primitive/rudimentary networks comprising concentrations of artisan skills, craft-specialisations and occupations.

These patterns forming rudimentary networks for each site are by no means diminutive. They form typical town-*kasba*-medium and small-village systems with their characteristic concentrations of rural engineering artisans and tiny processing units and unorganised workers. And with the *bazaars* at the *sundhais* or the *kasbas* facilitating the meeting of local needs largely through local production and petty trade, the network at each site covers taluka or tehsil size areas containing on an average populations of above two lakhs. These taluka size local economies in turn contiguously combine to form the zila level regional economies now also incorporating the production activity and scales centred around large towns or small cities.

Taking into account the local topological features these networks covercircular' areas in the plains and the plateaux, 'linear' economics following the ranges/the valleys/the rivers and the 'prayags' in mountainous areas of the Himalayas or the Western Ghats, or along the coasts forming a beaded necklace of settlements engaged in fishing and exchanging with the rice cultivating peasantry in their immediate vicinity.

It is such combination of scales at each of the sites comprising taluka level local economies that opens the possibility of consciously bringing together into a larger organisation, the small producers and workers etc. so as to provide the basis of assimilating without constraint any oi the agro-industrial technological flowsheets adapted and devolved in terms of these organised networks, or of incorporating tiny scale technology into combinations where economies of scale are visible with regard to shared operations or provision of common resources.

This Ability to Network is the Foundation for Empowerment.

The process of empowerment of the people is critically co-joined with identification and development to field activists working with agricultural labour, artisans, small peasants, tribal workers, fisherfolk etc. They have emerged as a result of patriotic youth movements, movements and field workers drawing inspiration from people like Gandhi, movements on such themes as environment, biodiversity conservation, wildlife conservation, peoples science movements etc. Industrial workers, equipped with the work culture of cooperation in production will be another major source of field activists. They will be engaged in the mobilisation of people, their organisation, operation of production units, upgradation of managerial and organisational capabilities of the workers in the networked units.

The field activists would need support from persons who can provide them the technological systems, training and backup they require for setting up their networked production organisation. Such a support will come from another group focusing on system design and development. Such a group will also develop the basic design for a region using secondary data and field data/experience by working closely with the field activists in their respective regions. These basic designs will then be adapted and detailed for any other site in the region. Such system groups may come from field activists, interested/motivated teachers and students from the local engineering colleges and other educational institutions and at the same time collaborate with individuals from national laboratories, research institutions who have developed lab scale technologies which are of interest and with experts in professional management and training.

The work of technology generation/improvement will generally be earned out in Institutes of Higher Education and research like IITs, CSIR Laboratories, Agricultural and Medical Institutes etc. Motivated/interested scientists/engineers in such institutions can act as the technology generators and problem solving groups for such a programme. Through interaction with system design groups and the field activists the technology generators can develop/improve technologies appropriate for the rural areas and suitable for the Rind of approach advocated above. Some of the sectors of technology needed to be taken up on priority basis are:

Waste Land Recovery and Watershed Management; Phyto-Chemical Processing Industry; Animal Husbandry; Dairying and Milk Product; Fruit and Vegetable Processing; Milling and Post Harvest Technology; Plant Based Health Systems and Referral Arrangements, Leather Technology and Products and Full Carcass Utilisation and Rural Engineering Sector; Energy Systems and Devices; Civil Engineering and Minor Works; Construction Technology and Housing; Pottery; Blacksmithy; Carpentry; Workshop and Foundry; Motor Repair and Maintenance; Electrical Shop; Electronic Assembly, Repair and Maintenance and further. On Job education with remote Access to Resource Materials and Expertise and Digital Computation and Communication Network.

In every one of the above sectors there will have to be a specific focus on women and child care, and on the family welfare.

Viewed in terms of the scientific/technological disciplines, one has need for the following illustrative disciplines:

Food Technology and Phyto Chemical Engineering; Agricultural Sciences; Forestry; Botany; Animal Sciences and Zoology; Chemical Engineering; Civil Engineering; Construction Technology; Mechanical Engineering; Electrical Engineering; Software Engineering and Information Technology; Electronics Engineering; Documentation and Accounting and Analysis.

It is unrealistic, however, to expect that every site shall have its unique-technology with nothing common with any other site. For one, the existence of typical patterns common to each of the regions/zones and bearing affinity across regions/zones leads one to more reasonably expect that technology packages typical to a geographical or agro-climatic region may be developed with options also developed and available for the technology package to be further detailed and adapted to the specific local conditions.

In other words any pilot scale programme will have to develop technology (with option) through action research for each of the geographical and agro-climatic regions and its sub-regions, by working at a sufficiently representative number of sample sites.

Methodology

In its broadest overview the evolving methodology based on experiences at I.I.T., Delhi and the work, and hence also the plan of work, can be parcelled in terms of [(3 + 1) + 1] major modules.

The "3" major modules relate to the: Field Activists Groups 3 Modules:

(i) Orientation and training, Field Investigation and technology Specification; (ii) Mobilisation of the People, Adaptive RDD and Formation of a skeletal production Network, and Development of the production Network to full Viability and Version.

Needless to say, the work in each module prepares the way for work in the next module and so forms a relay.

The 4th major module pertains to the Systems Design and Development Group which forms a continuing effort, but structured by virtue of its interlocking with the above "3" major field related modules.

The System Design and Development Group will interface with laboratories etc. through: Technology Generation Projects which represent the 5th module. This activity forms another band of continuing work, and again structured in terms of field level identification, trial and testing, pilot runs, feedbacking etc.

In the first module, while the field investigation has necessarily to be decentralised to the ground level, the orientation, training, opportunity analysis and technology specification can be nodalised with the initiating lead systems Design and Development Group playing the nodalising role and arranging also for interaction and dialogue with technology generators/ specialists.

In the second and third module the major work will be at site level with the members from the Systems Design and Development Group and technology specialists providing on-site, training, adaptive RDD and consultancy. There would be nonetheless, some elements of nodalisation, for example, for initial training for new/improved technology on pilot scale units with technology specialists and for subsequent monitoring and sharing of experience to make the field work more effective.

Orientation and Training, Field Investigations and Technology Specification

This module itself comprises sub-modules as below:

Orientation and Field Investigations, Opportunity Analysis and Technology specification; Preparation of "Adaptive RDD and Operational Programme"; and as a parallel activity, and Identification of Personnel and Locations for setting-up the skeletal production Network.

The nodalised orientation and training can utilise the census data on workers in the tehsil-wise village directories, to bring out the contours of the local economies in terms of the small nodal town (N) in the area, its connected big *bazaar/kasbaa* village (P), and their connected clusters of 2-3 medium (M) and small (S) villages with their concentrations of daily wage labour and peasantry.

This analysis and orientation makes the field activists recall their detailed familiarity with the geography and the economic activities imbueing their areas, and then to look at in a new way, in a different way so that they begin to perceive through a welter of facts, the prevalent system and the rationale underlying its occurrence and thereby change their own perceptions.

Another important feature is used for planning the survey in a cost-effective way. In obtaining information field activists are encouraged to hold compact group discussions and conversation comprising persons of the same occupation or with their buyers/sellers also present. The compact group discussions—interactive and lively—rapidly provide the cross-checks, the averages and the exceptions and the systematic buyer/seller connections. Field activists are encouraged to think in terms of meaningful sampling. They are specifically discouraged from embarking on detailed door-to-door survey of the census type. During the training, field activists are acquainted with such tools as 'production flowsheets' with details on input/output materials, tools, equipment etc., and with 'break-even' analysis to determine project/investment viability and the specific data and data-types required to arrive at the appropriate decisions. And, as the project/investments pertain to primary production, the data sets for break-even also include pertinent sketches/maps. These sketches/maps, sampled data, as well as systems diagrams bring out the area-wise input/output linkaging with other sectors. The field activists are encouraged to become thoroughly acquainted with data details sought, so that they can enter into a dialogue/discussion without intrusion of questionnaires. At an early stage the data and field work is nodally cross-checked to ensure that the methodology and concepts have been internalised. The maps/sketches, the systems/diagrams displaying input/output relations among and between the different buyers and sellers, and the production flowsheets describing the techniques in practice are analysed in detail with the field investigators by the systems-design developers along with the specialists involved. This is done nodally and intensively so that the opportunities for improvement and the type of systematic solutions with their multi-sectoral linkages, as also their likely impacts, start emerging. After a quick filling in of data gaps the report writing and the proposal for investment gets underway. The resulting document provides the adaptive RDD and Operational Programme for each of the participant sites.

Mobilisation of the People, Adaptive RDD and Formation of a Skeletal Production Network

The following are the sub-modules: Training of field activities and *'ustads'* on Pilot scale facilities and Mobilisation of People and Adaptive RDD on Site, and in parallel, Formation of Consultative Arrangements with the Small Peasants, and Organisation of Rural Workers and Artisans for the new Technological Initiatives, followed by: Start-up and Operation of the Skeletal Production Network, Proposal for Term Loan from Financial Institutions.

The training of field activists and the *ustads* who have been identified to set-up and run the units and to train the additional workers will be on the pilot-scale facilities of the technology specialists in the case of new/improved versions of technology. Once field groups have adapted and assimilated the technology the Systems groups will make arrangements with them to train new field activists from/for other sites. This measure will obviously augment the training facilities. It will also help relieve the pressure on technology specialists allowing them to concentrate on feedback from the systems group to develop newer/ improved versions of the technology.

The adaptive RDD is then an appropriate phase around which to design the mobilisation of people and engage in science-communication with the workers, the small producer suppliers/ buyers of input-output materials from and around the cluster of M/S villages or at the time of meals and weekly-bazaars where larger numbers may be made aware. The adaptive RDD and detailing to local conditions reopens in a way the entire gamut of issues. It subjects the proposed solution, to a concrete evaluation, to modification, and if need be, to a radical respecification—what specific local engineering materials to use as equivalents, or which fuel would suit with desired characteristics, the specificity of the local inputs, the ambient conditions, the specific practices and resource contexts, etc. the payback periods and interest rates on loans mobilised by the field activists from the small producers who would be beneficiaries in terms of the main operations once it got moving.

In parallel to the adaptive RDD, or as a prelude to it, or even in the extreme where suitable technology is not readily available. It has been felt necessary and significant to trigger the mobilisation and the organisation nucleation process. The idea is to create some form of consultative participative arrangement which facilitates and enables bulk-procurement and, or a pooling of small savings for contingent consumption credit needs, and eventually to create a debenture fund for investments by the Field Groups.

The systemic shift in the local economy involving the mobilisation of the buyers, the processors, the sellers, and the setting up of the allied multi-occupational consultative arrangements by the field activists, who themselves constitute a new value adding occupation, starts creating the first elements of a multi-sectors area-based planning to enable the specification and development of the technology system and the production planning for structuring the local economy in favour of local value-addition by intermeshing production units/networks employing the rural workers.

It however, needs to be reiterated that the development of consultative mechanisms and the shift of economic relations on the basis of existing small producers or handtool techniques, can only serve as a temporary prelude to and not as a substitute to, the adaptive RDD for the development and deployment of the new technological systems for production networking.

The interactive form of adaptive RDD and accompanied by an emergence of some degree of consultative mechanisms with the small producers and rural workers paves the way for operating the skeletal production network using the facilities set-up during the adaptation and detailing work. Once adequate experience has been gained on managing and organising the network and the functioning of technology the levels of risk are expected to have reduced to an acceptable 'routine' level. At these acceptable risk levels a loan from financial institutions (e.g., NABARD, SIDBI) may be sought with a conscious deliberated phasing out of the grant element from the normal entrepreneurial operations, but persisting for provision of feedback and for in-situ innovation towards the next version/generation of technology.

Development of the Production Network to Full Viability and the Next Vision

This phase gets operational on receiving loans from the financial institutions. The details and definitions of projections of the sub-modules will become clear from the experience gained in the previous module. One can broadly anticipate that for each of the sub-areas comprising the local economy the field activists trained on job will undertake the formation of consultative mechanisms for these sub-areas and create the skeletal networks in the sub-areas and progress on to their integration to form developed networks resulting in full NBMS production network for a given sector.

To achieve this expansion on job training will be in full swing and development of intra-organisational participative management initiated with the workers. The facilities for Adaptive RDD will be now used for continuing technological improvements and advanced training accompanying the on-job one.

10

Entrepreneurship Development– New Approaches and Opportunities in Rural Setting

Dr. S.P. Mishra

The Underlying Concept

The growing awareness of the need for and urgency of, building "entrepreneurs" for the attainment of accelerated and self-sustained economic growth could be ascribed to two factors, first, the belief is gaining ground that economic growth in the advanced countries appears to be attributable to entrepreneurial awareness in the community rather than to capital. In fact, several of the empirical studies recently conducted show that entrepreneurs or the 'human capital' has grown in the western societies at much faster rate than conventional capital and has made a much larger contribution to economic growth that non-human capital. Second, investment in human resources has directly contributed to economic development and growth, by promoting the knowledge and application of science and technology to production process, developing innovations and research, training the workers in different technical skills needed for modern production and building up of the right type of attitude, values and interests conducive to right output. The fundamental problem in developing countries is, therefore, not so much the creation of wealth but the capacities to create wealth towards strengthening widening and improving the absorptive capacity of the country.

In other words, entrepreneurship development is an approach of developing human resources. It is concerned with the growth and development of people towards high level of competency, creativity and fulfillment. This approach helps people to grow in self control, responsibility and other abilities and then try to create a climate in which all client may contribute to the limits of their improved abilities. It is assumed that expanded capabilities and opportunities for people will lead directly to improvement in operating effectiveness. Thus, Entrepreneurship Development as an approach does not confine itself in setting up enterprises but surpasses this limit in creating conducive climate for optimum utilization of limited and scattered resources and making people functional in all walks of life.

Entrepreneurship is the creative response to an environment that combines innovativeness, readiness to take risk, sensing opportunities, heightened initiative, standard of excellence, persistence in achieving the goal, positive orientation to problem solving and constant striving for growth and excellence. When all those attributes are developed in one person, the person can be found in any field of activity such as industry, business, education, public or professional bodies etc. But wherever they go they create a landmark, turn the direction of the side and attain heights.

The Experiences

The concept of entrepreneurship has undergone remarkable changes during the recent decades. Professionals of different disciplines contributed largely to the growth and maturity of the concept and application for economic growth of the society. During the last three decades, academicians and practitioners contributed significantly ' theorizing and applying entrepreneurship concept which resulted into few important generalizations:

- There is positive linkage between entrepreneurship and economic prosperity.
- Entrepreneuryship is instrumental in achieving progress all walks of life.

- Entrepreneurship can be developed through planned training intervention/efforts.
- Entrepreneurship is a process consisting of stimulatory, support and sustaining activities.
- It requires to be viewed in terms of integration of activities and coordination among different departments.
- Entrepreneurial culture need to be in built mechanism of development process.

These findings helped acceptance of different approaches to develop entrepreneurship. One of them is the development of human factor–the entrepreneur himself, another major factor is the development of environment through which entrepreneurial activities can flourish and grow. The human factor refers to the values, attitude, desire and motivation of individual, his capability to perceive the environmental changes and opportunities as well as his ability to solve the problem which entrepreneur is likely to face. All these qualities differ on the average from country to country or culture to culture and these differences are related to economic growth. These qualities known to be entrepreneurial qualities, are not inherited by a person rather acquired in the process of socialization in a particular culture. Emergence, growth and development of entrepreneurial qualities takes root in childhood and continue to get sharpened with conscious attempt.

Many attempts in promoting entrepreneurship in different countries have yielded variety of experiences. Some were highly successful other disappointing and perhaps considerable success. Analyzing these past experiences mainly from the perspective of the client system the whole process of entrepreneurship development can be seen as a process involved in generating economic activities, multiplying such activities and finally ensuring its continuance.

Recent Approaches for Entrepreneurship Training Programme

With the increased emphasis of developing entrepreneurs, witnessed a phenomenon change has been witnessed in its approach to create entrepreneurial awareness and generate entrepreneurship in the society. What appears to be most significant is the realization that no approach can meet all requirements in dealing with diverse population scattered over widely dispersed area of any country. Consequently, different types of entrepreneurship development training programme took a shape in making it as an effective preposition for industrial development. Recent approaches to entrepreneurship training may be seen as:

- Target oriented
- Location specific
- Product and process oriented

Target Oriented

Such EDPs are directed and planned for a specific group which has distinguishing features as compared to other groups. It is because of the fact that target audience vary in terms of their background, experiences, training and exposure to business world. Moreover, the size of the proposed enterprise may also vary from one target group to others. These variations demand for the matching training inputs with varying degree of intensity to bring these entrepreneurs to the threshold from where they can launch their enterprise and be able to manage them successfully. For example, a group of MBA entrepreneurs may require less intensity on management skill development input as compared to other aspects. Similarly, for a group of entrepreneurs who want to setup considerably larger units may require high intensity on management skills rather than those who are going to set up comparatively smaller units. For this purpose, various target groups may be identified like (i) General Entrepreneurs, (ii) Science and Technology Entrepreneurs, (iii) Women Entrepreneurs, (iv) Educated unemployed entrepreneurs etc.

Location Specific

In order to reduce regional imbalances and the imbalances that may exist between progressive and non-progressive areas, it is of considerable importance to plan EDPs depending on the characteristics of the area itself. Location in terms of urbanization, concentration of specific target group in a particular area etc. may be considered under such specific EDPs.

Product and Process Oriented EDPs

The late eighties is experiencing a shift from Target Oriented EDP. Such EDPs are organized for a group of prospective entrepreneurs who opt for enterprises having set product line or process such as plastic or electronics or construction materials of food technology etc. It seems to have a great future, but some of the efforts made recently were found suffering from the lack of appreciation of the role played by inputs, other than product and process orientation. In the absence of such comprehensive programme, the acquired knowledge and skills about product and process will not automatically result in development of entrepreneurial quality, competence for enterprise launching and ability to manage which are crucial to start and ensure success.

Entrepreneurship Development through Planned Training Intervention

Training in entrepreneurship provides an impetus to the potential and budding entrepreneurs to acquire a new identity about himself. This is perceived as an approach towards transforming people which serves the purpose of making people aware about their own identity, help them accept a new identity and finally establish such identity for entrepreneurial pursuit. In order to take up such task of transformation we can find entrepreneurship training serving the purpose of stimulation, orientation, preparation and implementation in a sequential order. In order to apply the model, one requires an understanding about approaches to entrepreneurship training programme.

Entrepreneurial Approach in Training

The entrepreneurship training differs with other type of training in terms of its nature, scope, result, target group, post-training activities etc. Promoter of training to the small business requires to undertake a business like approach. This business like approach to training must also, however, be entrepreneurial in delivery and marketing. Following are the salient points of entrepreneurial approach to training:

Entrepreneurial Approach

(a) Major focus on process of delivery

(b) Ownership of learning by participant

(c) Trainer as fellows learner/facilitator

(d) Emphasis upon know-how and know-who

(e) Participants generating knowledge

(f) Sessions flexible and responsive to needs

(g) Learning objectives negotiated

(h) Mistakes to be learned from

(i) Emphasis upon practice

(j) Problem/multi-disciplinary

The entrepreneurial approach in training can be immense use in visualizing and immolating the training process in entrepreneurship.

Emerging Trends

The scope of wage employment is not only limited but decreasing very fast. Even the nature of wage-employment is undergoing drastic transformation from a security oriented job to performance based contractual assignment. Such a situation has virtually compelled People in general to think more for entrepreneurship/self-employment. In this process, an analysis of the situation indicate clearly that today's environment is pointing concern towards the aspects like:

- Highly competitive market
- Competition between public and private sector
- Quality consciousness and standardization of products
- Security/guarantee
- Fast changing technology
- Technological collaboration
- Liberalizaton in policy and action

- Health/Hygiene
- NRIs readiness to invest in India
- Privatization of public sector
- Increased export opportunities and mental readiness of other countries to buy products from India
- Emphasis on self-sustenance of supporting/promotional organization leading to better services from them.

Therefore, the environment in totality appears to be quite challenging. In such a direction the question arises as how to help people sense business opportunities which fulfills the global expectations in general and local needs in particular? For this an attempt is made here list out possible models in operation in one hand and emerging demands on the other hand.

Models

A few models in operation are as follows:

(a) Market Driven Model

Demand estimation and demand forecasting are the best ways through which demands of various products are reflected in the market. While the demand is estimated on the basis of the gap between the demand and supply, the forecasting of the demand is made on the basis of future projection of possible needs of the concerned products. Therefore, economic activity is generated related to this model to fulfil the requirement of the market.

(b) Need Based Specific Model

Generally, we find two types of needs, i.e., felt and unfelt needs. It is possible understand the felt need on the basic requirements and the gap but it seems difficult to visualize the unfelt needs, i.e., need below the surface. In other words, unfelt need to indicate the need in reality but in latest form. This is one area on the basis of which various products find place in the market. Related to this is another concept which is known as need creation in the society. There are number of examples where in different products have been introduced through this process in a particular social system like emergence of fast food in traditionally strong society in African and South Asian countries.

(c) Uniqueness Model

A variety of products have been introduced in the market by projecting unique features of particular product for specified segment of society. This may be seen quite clearly in automobile industry, i.e., cars, two wheelers etc. Based on these models and some more conceptual models already in operation, the following emerging demands have come in to focus.

Emerging Demands

1. Focussed on Ecofriendliness

- Eco-friendly food/vegetables/crops
- Eco-friendly cotton/thread/cloth
- Eco-friendly colour
- Eco-friendly house material etc.

2. Cow Based

- Cow Milk — Milk products
- Cow Dung — Dhoop Bathi, Agarbathi
- Cow Urine — Insecticides etc.

3. Herbal Products

- Cultivation — Medicine
- Processing — Cosmetics

4. Non-Conventional Energy

- Waste Conversion
- Energy Supply
- Energy Production
- Hardware—Solar Devices

5. IT Enabled Services

- IT Clinics (Doctor)
- IT Tutorials (virtual classrooms)

- E-commerce (Vegetables, bending)
- Shopping Mall
- Communication–Telephone, Email, Voice mail etc
- Networking
- Processing-accounting, billing etc.
- Internet
- Simulated games-exercises
- E-entertainment
- E-designing, E-Training, E-Education etc.

6. Relax Points

- Yoga/Meditation
- Spiritual camp/discoursed
- Stress management centres

7. Household Service Points

- Electricity, Telephone, House Tax, LIC premium etc.

8. Counselling Clinics

- Career Counselling
- Conflict Resolution
- Clinical Counselling
- Behavioural Therapy

9. Research and Development Based Enterprises

- R&D Centres—experiments at IIT Kharagpur

10. Agriculture Based

- Scientific cultivation of crops, cash crops, crop rotation, crop intensity, multiple cropping etc.
- Processing of cereals and spices
- Vermi Compost
- Vermi Gel/Liquid

11. Input and Output Based Enterprise

- Dal Making
- Rice processing/screening
- Vegetable grading/packaging

12. Specialized Service Enterprises

- Tours and Travels
- Repairing
- Home/Door Services

13. Skill Based

- Carpentry — Mechanic
- Blacksmithy — Repairing and Maintenance of
- Goldsmithy — Agricultural/implement/equipment, pumps, household appliances etc.

14. Packaging and Marketing Based Enterprises

15. Enterprises towards reviving traditional handicrafts, design and products

Conclusions and Suggestions

In this context, it seems important to take advantages of the earlier experiences and research findings which may be summarized as follows:

- Promotion and development of entrepreneurship is more developmental in nature and is not administrative or target-oriented.
- Entrepreneurship development is not a programme but a process whereby people in general are stimulated, supported and their enterprising activities are sustained.
- Accumulated experience has brought maturity in approaches and strategy of developing entrepreneurship.
- Experience suggests that entrepreneurs could be developed through a well-planned integrated approach.
- A set of competencies is already identified to help people become entrepreneurs. These competencies could be developed through training intervention.

- Competencies could be broadly grouped as:
  - – Behavioural
  - – Enterprise launching and resourcing
  - – Enterprise management
- New economic opportunities are identified and available which may be integrated in entrepreneurship development programmes.
- Special attention may be made to promote "marketing entrepreneur" and R & D entrepreneur.
- Emergence of women entrepreneur through women empowerment seems to be an established fact.
- Micro enterprise creation and development through micro-finance has shown a very significant result.

These findings could be used to promote entrepreneurship professionally in the country.

11

Women and Appropriate Technology in Rural Industrialisation

G.K. Ghosh

The topic here covers many questions at one go. It covers women development, rural development and technology appropriate for such a growth. While others are being covered in following paragraphs in a phased manner it is important to indicate importance of women development in rural sector. According to statistics broadly available, it is said atrocities on women development in wheat growing areas are much higher than in rice growing areas, at least in rural areas. It is so because wheat cultivation has been modernized hence relevance of women in economic cycle has been reduced, which is not so in rice growing areas. In other words involvement of women in rural development shall ensure better law and order situation too.

When author was discussing about this topic with one of his modern urban friend, his friend expressed his surprise saying what role women can play in industry which is a domain of men. Author do not blame his friend for this ignorance since his friend works in a large industrial house and moves in various chambers of commerce or elite clubs. He could create ideas of world within sphere of his movement with privileged few who enjoy about 95 per cent of the wealth with their population less than 1 per cent. In fact exploitation in human civilization started with exploitation of women before other form of exploitation like economic or political exploitation taken its root. It is very much true even today as otherwise why there was volleys of protests from various of fundamentalist organizations when

Tasleema Nasreen an educated young women raised certain social question through her creative literary works.

If we study the background our various industries in details including the evolutions, we can notice that number of industries were originally the exclusive domain of women but were snatched away from them by men after mechanization, automation, etc. This change over neither done good to our society in any form. First of all in case of many of these industries, e.g., food products etc., is still known to women depending on particular area of inhabitance. When men adopt this process for implementation of industry they have to go for various non-productive investments like consultancy or technological transfer otherwise not necessary for women. Secondly, women used to adopt in decentralized way in rural areas hence waste out of the process used to go back as nutriants. As men took over and mechanized them, it had to be centralized industry hence the process waste became urban refuge. Energy input through labour and brain from women was discontinued hence we had to go for fossil fuel or electricity to run these centralized industries keeping available energy idle. Lastly by making women irrelevant to economic activities, atrocities on women increased at galloping rate causing law and order problems, thus increasing budgets for police, increase the taxes to meet this budgetary expenditure and finally ensuring inflation.

When programme of industry has to be thought of one has to decide various types of industries. Broadly we may categories various industries as (a) Agro based (b) Mineral based (c) Polymer based (d) Service Industry (e) Mechanical (f) Electrical (g) Electronics and so on. Agro based industry, however, was original of all industries. It may be categorized further as (a) edible (b) non-edible. All industry that derives raw material from botanical kingdom comes under this category. If the end product is edible in nature it is edible agro based industry while if it gives non-edible end product, it may be put into second group.

Edible agro based industry traditionally belong to women till mechanization over run the original tradition. Entire processed food industry, processing of foods like cereals, pulses, etc. extraction of oil from seeds, producing various edible medicines from plant kingdom comes under this category belong to women a tradition still continues in many of the third world villages.

In fact, women have much more flexibility to switch over from one raw material to other if situation arises at amazing speed hence processed food industry can go on if it is run by women who have the right in this type. Let us begin our journey from Nigeria, the well known African country. Duddawa (also known as dawadawa, Soumbala and local magi) is widely used condiment in West African cooking. It is traditionally made from locust beans from the Parkia biglobosa tree. Probably because daddawa is an indigenous food, sold at village markets and traded via traditional rather than modern commercial channels, it had received next to no attention in official statistics. Similarly, the locust bean tree has been neglected in agricultural/forestry research and extension. Nevertheless, amazing changes are occurring in the daddawa business in central Nigeria.

Within the last few years, daddawa is increasingly made from soyabeans. This innovation can be credited to the ingenuity of Nigerian farm women who developed a soyabean processing technique which requires much less time and labour than locust-processing. The greatest saving is the reduced cooking time (6 instead of 24 hours) and thus the reduced need for cooking water and fuel, furthermore, the women can grow the soyabeans on their own fields instead of having to buy locust beans or tree harvesting rights. The trees are usually owned by men.

Information about soyabean growing and processing has spread from women to woman, not only within the indigenous Kaje ethnic group. Also Fulani families who have settled in the area have benefited from the Kaje women's extension activities. Kaje women gave Fulani women some seed to sow in their kitchen gardens, advised them about cropping techniques and showed them how to ferment the beans to make daddawa for home consumption. These activities remain largely unseen by the outsiders who consider themselves to be the major (or sole) catalysts for development. As it stands today much of the daddawas old in Nigerian market is now made of soyabean, a result of technology development by women.

Traditionally, women used to develop their processed food from local plants but ready to switch over if need arises but maintaining a self-reliant economy as mentioned above. Indigenous plants provide diverse and nutritious food and may be vital during hungry seasons and famines. They can also provide many useful non-food products and source of income particularly for women. As they often grow on marginal land or form part of multi-cropping systems, exploiting indigenous plants increases land-use efficiency.

The major commercial trees of the tropics and sub-tropics, e.g., coffee, tea, cocao, coconut, oil plam, rubber and citrus, are well known. Lesser known are the products of indigenous economic trees such as Shea butter [Butyrospermum Parkii Syn. Vitellaria paradoxa), Marula plus (Sclerocarya birrea), African locust bean (Parkia biglobosa) and local magi made from it; leaves of baobab and bitter leaf which are used as vegetables; spices from the Akee apple, red Kapok and Calabash nutmeg and medicinal ingredients for treating humans, animals and plants.

Many products from indigenous trees are valuable local foods in terms of both quality and quantity. For example, one mature locust bean tree supplies about 40 kg. of seeds in an average year. When, fermented, these have the same nutritional value as 50 chickens and provide considerable amounts of vitamins and minerals in a period when other foods are often scarce. In a hungry season, the products of local trees and shrubs may be the basis of survival for many farm families. Since only local women know-how to prepare various dishes out of these plants, their knowledge may be utilized to produce new type of dishes and packed appropriately for commercial use to ensure additional income to these rural women entrepreneurs.

Indigenous economic trees are seldom as "Wild" as they are perceived to be by outsiders. They are often integral part of local farming systems; they may be protected and managed or even deliberately planted by farmers. Faidherbia albida in Africa and Prosopis cineraria in India are examples of indigenous trees that play a central role in land-use system in the semi-arid tropics. Such trees are often vital for the sustainability of farming and their disappearance may lead to ecological imbalance and social and economic tension. Especially poorer farmers and women who derive an essential part of their food and/or income from trees and shrubs suffer when these become scarce. Table 1 gives source of income for men, women and poor women in Uttar Pradesh, India.

Table 11.1
Source of Income for Men, Women and Poor Women in Uttar Pradesh, India (Per cent)

Income	*Forest and Common Land*	*Cropland*	*Off Farm Activities*	*Total*
Men	13	28	59	100
Women	33	35	32	100
Poor Women	45	34	21	100

The above table shows men depend less on cropland produce and much less on forest or common land produce but depend more on off farm activities. While on the other hand women depend on crop land and forest or common land produce rather than off farm or non-farm sector. In fact poorer women depend more on forest or common land produce. These figures are eye opener for all of us to decide future modalities in women programme.

Drought prone Kalahandi district of Orissa, India could sustain itself without additional external aid till 1965 but today drought had to be fought with aid from outside. This is so because the local plants which sustained them has vanished due to deforestation or due to commercial plantation under new afforestation programme which did not do any good. Efforts were made to provide employment to women and men of Kalahandi under special programme sponsored by Khadi and Village Industries Commission and District Rural Development Agency (DRDA) but it did not take off well since women is general were not ready to spin yarn under a common shed but preferred to depend more on minor forest produce even on a product as low cost as fire wood.

In fact all over third world nations, there is a special relationship between trees and women. This aspect was never given due attention in any stage. Often, insufficient attention is given by development planners to the problems of involving women in tree planting. Enabling women to derive benefits from tree resources, e.g., by involving them in the cash economy of tree cropping, is likely to prove a rewarding strategy in tree planting.

Many processed food products known to a particular locality or country could be universalized. These processes were originally known to women and developed by women but as soon as these products were commercialized and finally mechanized the process and product development were taken over by men. On the other hand women could adopt and learn many processes unknown to them but were imported but it is certain it shall remain with them till manufacturing remain small scale but as it shall grow it shall go to the kingdom of men. But still there are many more processed food which remain traditional or even after improvised remained with women.

Mongoloid culture still known as bamboo culture. Bamboo provides many necessary materials to vast Mongoloid population like food, housing and many more. Vast Mongoloid population of China, Japan, South-East Asia, North-Eastern India, Himalayan belt besides adjoining non-Mongoloids of Assam, Bangladesh and West Bengal still depend considerably on bamboo. Bamboo shoot is one of their important food which they cook as side dishes or pickles. Bamboo shoot pickles are gaining popularly even though it remained traditional so long. In Nagaland various women organizations with active support from government are packing and selling bamboo shoot pickles which are being sold through various commercial channels. This trend is visible in various other regions too, but it is facing threat from various organized paper mills which are eating away vast bamboo wealth at very fast rate. Similarly, fish eaters produced by burying fish with water underground for several months is a popular dish for taste amongst people living from Comilla, Sylhet of Bangladesh upto various far eastern countries are yet to be sold through usual network of commerce but yet is a popular item. Its production can ensure income to women of the region and can be popularized. Similarly fish dried in such sold as dry fish is very popular throughout the Orient and in part of India. It is a way to preserve fish for month together and sold through traditional markets. While men folk catch fish, women folk dry them under Sun and sell them thus ensuring income. This can be organized as part of women development programme.

Cocoa is now a cash crop that has high demand round the globe. Thanks to a number of multinationals that exploited this crop from third world to ensure high profit for them while cocoa otherwise available to commoners earlier has become dearer. In Ghana or adjoining African countries where cocoa traditionally grew wild used to prepare various nutritious traditional cakes from traditional wild cocoa which they cannot afford now. In turn we are consuming high cost, low nutrition chocolates from cocoa from multinationals ensuring low income to cocoa farmers and throwing out women out of employment who used to produce traditional cocoa cakes. Though Belgium and Switzerland now famous for their chocolate clusters, they draw raw materials from Africa, symbolises exploitation.

Large baobab trees grows wild in Africa and now found in Western Coast of India after the Muslim traders introduced them. It has various utilities both for ritual as well as industrial in Africa including preserving dead bodies for several days inside in the hole of its trunk. This large tree, that lives long often shelters few families inside its trunk. The pulp of its fruit, diluted with water and sugar consumed in Africa as cold drinks from this fruit often sold in weekly markets in Africa but it is now facing stiff competition from branded cold drinks which has money power and power of media. This is true in case of all types of traditional cold drinks, freshly prepared and sold by women are loosing its marketing base against cola lobby of multinationals. Third World is now buying sugar solutions or flavoured water at high cost rather than cheaper more nutritious fresh fruit drinks. Even national governments are surrendering to the pressure to the cola lobby obviously due to pressure from white capitalist governments. Finally we are throwing women out of employment and paying high cost for less nutrition and harmful drinks.

There are a number of fruits which had their origin in Third world countries. Mango known for its taste and nutrition is a typical example. Indian sub-continent is known for quality mango, each variety of, has sits own aroma, taste and flavour. Indian is known for its mango based industry. Unripe mango is used mostly for pickles, sweet and sour drinks and souring agent. Even crushed unripe mango paste is a favourite side dish in many Indian families. Ripe mango is consumed as fruit, as fruit juice, as a part of fruit salad besides preparing various sweet dishes. Mango pulp dried as mango jelly is sold as traditional processed food, preserved and consumed various ways, either as it is like dry fruit or cooked with various dishes. They are dried soft semi-dry or hard fully dried depending on the demand. This is a well known cottage industry in India employing women. But we have invited various multinationals to kill this tradition who buy the entire mango for their process houses either for export or sold at very high price to us and it is very clear where we are leading to.

Pickle industry similarly remained as tradition in Azamgarh in Uttar Pradesh, various areas of adjoining districts upto Varanasi, various places in Andhra Pradesh, Tamil Nadu and Maharashtra while Bengal and Bangladesh was known for sweeter pickles, like morobba, Kasundi etc. In these areas it was rather a home industry and women knew how to produce various types of pickles from various vegetables depending upon seasonal or regional availability. This is one industry that ensures 100 per cent employment for women. Fortunately multinationals are yet to extend their hand towards pickle industry although organized sector has already started pickle industry in a big way, and if this tendency continues in a big way pickles shall be produced in organized factory this home industry shall close down throwing women out of their livelihood. A few voluntary organization all over the country are trying to protect this traditional sector too. For example, Diamond Silk Khadi Society in Malda in West Bengal or Cuttack Zila Mahila Samiti in Cuttack (Orissa) are trying their best.

Fruit products like jam, jelly, squash, sauce etc. are not Indian or third world concept but it is an example of technology transfer from various other countries including West but adopted well due to simple technology. They are now being produced as home industry in India as well as many third world countries. Many of these products are now being produced with various locally available fruits and vegetables. For example, pickles or jelly produced in China from Chinese rose (Hibiscus Sino-chinesis) known as sacred Jabakusum India, is produced both in traditional as well as new methods are known both for nutrition as well as taste. The flower reported to yield all the Vitamins of Vitamin 'B' group. Similarly jam or jelly from guava fruit or mango is popular in India. Products from citrus fruits or pineapple or apple are well known. But these products are now being produced by organized sector even the multinationals hence it no longer belong to women. What is more alarming being these organizations have switched over to synthetic jelly adding flavour and colour which is cheaper than those produced from fruit pulp hence are accepted well by the consumers due to aggressive marketing strategies. This recent trend on the one hand thrown women and people below poverty line out of employment while reducing intake of nutrition. In spite of all attempt made by NAFED or voluntary sector it failed to make any dent against powerful organized sector since people in general are addicted to publicity.

Cereals and pulses are our main foods which are to be processed even for normal consumption. For example, paddy cannot be consumed but has to be consumed as rice after dehusking and polishing. Hand pounding of rice or even removal of paddy from straw was traditionally done by women. As power driven dehuskers or polishers or threshing machineries were introduced men took over but some extent women also continued the operation. But large scale rice mills were introduced it was social, economical and human disaster. Rice mills were mostly owned by hoarders or profiteers. In one hand they exploited the farmers while on the other hand increased the price of rice by hoarding ensuring women out of employment. It is true in case of pulses also through not to the extent of rice. Some local technology of processing pulses were forgotton which made many of the pulses useless. For example, Khesri daal or lintel normally considered as harmful, is popular in Orissa since Oriya people had local technology to process this pulse appropriately which modern technology cannot. Hence as popular pulses has become a goods unfit for human consumption. Hence modern technology has reduced the source of food.

Traditionally cereals and pulses were used for producing various processed food items, mostly employing women. Tribals consume rice beer, brewed by women. This low alcohol rice beer is rather a healthy drink but modern thinking and imposition of various other culture has encouraged the tribes to switch over to unhealthy drinks like tea, coffee or stronger alcohols like whisky, rum etc. causing deterioration of their traditional drinks prepared with maize, job tear, etc. were tastier too. Tribal women no longer brew their traditional healthy drink which has vanished along with their social feasts, festivals and gatherings creating a society of violence, envy and tension. From pulses our women traditionally used to produce various items popular even today, "Papad" prepared by rolling the dough of powdered pulses is extremely popular. Technology is simple and known to women. Sri Mahila Griha Udyog "Lijjat Papad" is a voluntary organization run by women having its head office at Bombay and having branches all over India. They produce high quality papad at low technology employing women living below poverty line. It is a well known organization and every worker be Rukmini Tai in Bombay or Preeti bahen and Mistry bahen in Calcutta works for the cause. They have recently introduced detergent cakes and powder employing women. Sunderban Khadi and Village Industrial Society, a women organization having its headquarter at Canning, West Bengal has also started production of Papad and Badi employing women. Abhoy Ashram another organization of West Bengal is also producing Papad employing women. Badi in fact is also produced traditionally by sun drying lumps of slurry from pulses. Type of papad again varies considerably from region to region. Types of papad produced by women of South Orissa and North Andhra Pradesh is with puffed rice powder or gingerly seeds which has typical taste. In Orissa "Badi" produced with seeds of pumpkin is also popular.

Puffed rice (Moorhi or Moora), flattened rice (Poha or Chuda or Chida), Puffed rice (Khai) are the items produced at home by women which is slowly being mechanized. This should continue as home industry ensuring womens participation. From these items sweet dishes are also produced with either molasses or sugar solution. These dishes are commonly known as "Moa" in Eastern India, Moa is also produced with gingerly seeds.

Laddu or Ladu is another similar sweet dish produced out of dry powder of pulses with sugar or with crushed powder of coconut pulp or with various types of dried sap of various plant. Paluo, Ladduone such produced from dried sap of a wild plant found in Orissa. Paluo sap is also considered as highly medicinal. All these sweets may be produced and sold. In fact various types of food items out of cereals and pulses are produced all over Third World mostly produced by women traditionally.

In Far Eastern countries similarly have the tradition of producing various such items out of cereals and pulses. Chowminh or noodles are the items produced out of cereals has monopoly of China and Mongoloid world. Simei noodles produced out of wheat flour is the delicacy of Muslim world out of which porridge is prepared with milk and sugar. Rengma Nagas of Nagaland produce a sun dried dry sweet out of goosc berry and sugar; which may gain popularity if propagated. They also prepare squash out of various wild fruits which can gain popularity if appropriate preservative is added. In Bangladesh a sort of Sauce called Kasundi is prepared out of mustard seed which is very popular in both India and Bangladesh. Petha a sort of dry sweet prepared from pulp of a gourd is popular in Uttar Pradesh which can be prepared by women. It has its popularity all over India and can be preserved for a number of days. Though apple was introduced in India recently, women of Himachal Pradesh could produce dry apple chips in their own way which are now boing packed and sold. There is a saying in India that there are thirteen festivals in twelve months. This saying is applicable all over India irrespective of region, caste, creed or religious belief. There are folk dances, rituals and feasts connected with each festival. People enjoy feasts with traditional cakes prepared out of materials available around depending on season and region.

They also exchange gifts and send these cakes for social amity. These cakes are prepared by women at home using rice, molasses, coconut, pulses, vegetables, fruits or milk products. These cakes known as "Pitha" or "Pithey" are prepared either by baking or by frying or simply boiling as may be applicable for the type chosen. In all over India they are prepared by women. The cakes may be sweet, salty or sour depending on the type. For example, "Arisa Pitha" popular in Orissa is prepared by frying a dough made out of powdered rice and molasses. Kakara pitha on the other hand is prepared same way replacing rice with wheat flour. These are obviously sweet cakes. Backed rice cakes with appropriate stuffing is almost universal all over India wherever rice grows. Stuffing may be coconut pulp or pulses or any other gruel sweet, sour or salty. Chakuli pitha in Orissa or Soru Chakli is Bengal is prepared by spreading rice and pulses slurry and frying. Pati Sapta in Bengal on the other hand is prepared by folding Soru Chakli ansd stuffing appropriately. This is also popular in Assam where it is soaked in sugar solution. Various types of such cakes are also popular in South East. These traditions are slowly vanishing since these cakes have less commercial value not being supported by publicity. Few items produced out of ragi gruel by various tribes of Central and eastern India or with millets by tribes of North Eastern India Himalayan and Sub-Himalayan region are wonderful. Promotion of these items are self-reliant, cheap while ensuring nutrition. Instead of this healthy tradition we are switching over to food items produced out of materials brought from far flung places hence are costlier. Moreover since glamour is preferred, nutrition and quality of food takes back seat. A few, however, could gain popularity and publicity like "Podo pitha" Orissa which is produced by backing cheese in sugar solution.

It is generally believed that the technology of bakery is a Western concept adopted in Third World after colonisation. But if we study Third World tradition a little in details we can find that baking was very much a part of Third World method of cooking. In previous para discussions on Podo pitha or various other traditional cakes reveals this fact. Tanduri bread, or Tanduri chicken in Pakistan and Northern India is prepared by baking, the tradition which is still popular and is gaining popularity even in Western countries. "Idlee", the popular rice cake in Southern India is prepared by steam baking method. Various preparations both vegetarian as well as non-vegetarian are made by various tribes in Indian sub-continent by baking using various materials available locally like millet, job tear, ragi etc. However, bakery item from Western concept was accepted well in India though there was some initial resistance. Today breads, cakes, pastries and biscuits are accepted well as if it is the part of Indian tradition. With change in agricultural pattern raw materials are also available locally in India and Third World. These products can be produced at home as, home made bakery products but it could not take-off due to uneven competition from organized sector. These biscuits or cakes could have been organized easily as home products maintaining special delicacies employing women.

Various types of palm trees grows almost all over third world countries giving us various edible and non-edible products. It is said that there are not less than 300 species of palm that grows in various proportions in various tropics that in various Third World countries. Out of all oil palm, palmyara palm, date palm, talipat palm, coconut palm and Sago palm are fairly well known throughout the Third World. Oil palm known for its oil is a plant that grows naturally in plenty in South-East Asia in general, Indonesia and Malayasia in particular. Picking up matured oil yielding fruit has to be done from tall palm tree thus cannot be done by women. But those matured fruits that falls on the ground are usually picked up by women. Extraction of oil is done by organized oil mills now hence women has no part to play. Almost all palm trees yield asweet Sap or nector which is collected by men folk but is converted into sugar, candy or molasses by women at home. In Indian Sub-continent and Africa it is one of the popular home industry. In entire South India notably Kerala and Tamil Nadu these items are most popular sweetening agent used in various traditional sweets. The sap itself known as "Neera" in India is a popular drink all over Third

World. Problem with neera being it ferments very fast and has to be consumed within 2 to 3 hours of extraction. Neera, the nectar is the delicious drink extracted from sap of palm. Refreshing as it is, the drink has agreeable flavour, high nutritive value and medicinal properties. It is wholesome, cool and good for improvement of general healthy, specially as a supplement to those having iron and vitamin deficiency. Therefore, is builds the body, keeps the system cool, improves appetite and digestion. Neera having pH of 7.2, containing sugar, calcium, iron, phosphorus, thiamine, riboflavin, ascorbic acid, nicotinic acid, protein and sufficient calories. For male palmyra palm tree season for extraction is from December to February while for female it is February to March. In case of date palm the season is from October to March. This is more or less true for entire tropics. Hence on average we get six months for tapping neera.

Palm gur, also known as molasses or jaggery is prepared by boiling sweet juice of palm or Neera. The method of preparing improved variety of gur is quite interesting. Limed juice is collected, slightly heated, clarified by adding superphosphate of phosphoric acid, filtered to remove impurities. The clear juice is then further boiled upto 116°C to 118°C and moulded in different shapes and gur is ready for consumption. Date palm gur is more popular for its beautiful flavour. The composition of palm gur is given in Table 2.

Table 11.2
Composition of Palm Gur

Kind of Gur	*%*	*%*	*Reducing Sugar %*	*Fat %*	*Proteins %*	*Total minerals CO %*	*Calcium %*	*Phosphorous as PO %*
Palmyra Gur	8.61	72.86	1.66	0.19	1.04	3.15	0.861	0.052
Date Palmgur	9.16	72.01	1.48	0.26	1.46	2.60	0.363	0.52

In addition the palm gur also contains Thiamine (Vit. B), Reboflavin (Vit. B), Nicotinic acid (antipallagra Vitamin) and

Ascorbic acid (Vit. C). If the juice is clarified and boiled further upto crystallization, it gives palm sugar or palm candy. They are highly medicinal and known for appealing flavour. Entire process of production of gur, sugar or candy is done by women. In fact during winter both in West Bengal and Bangladesh sweets available in the market are made with date palm gur only. Gur made in Nadia district, Murshidabad district and Bardhaman district of West Bengal and Barishal district of Bangladesh are known for pleasant aroma.

Palm leaf from Palmyra tree was commonly used in India as a writing media in place of paper. Palm leaf is also used for manufacture of cracker and as packing material. It can be used now for manufacture of various fancy items employing women.

Extracting fibre from the palmyra tree is an ancient concept through tropics. The process of fibre extraction is done by a wooden hammer to separate the fibres. It is then passed through the comb fixed on a board to separate and remove additional fibre hairs. The fibre is then dried, sorted and bundled and is ready for export or other uses. A variety of articles like brushes, foot rugs and mats are prepared out of this fibre. Mid-rib of the palmyra leaf is called eark. This is also utilized separated for manufacturing several utility items. Yet there is another subsidiary industry in which fibre is extracted from leaves through carding machines. This fibre is then dried, coloured and used in weaving door mats, foot rugs and cot tapes. They are also directly twisted for rope making which has many uses in village. Date palm leaf mats traditionally used as beds in Orissa, particularly in tribal areas. Above all broom stick made from coconut mid-rib has universal use. While in entire third world tropics palm industry remained as unorganized traditional sector, in India after independence it was organized by Khadi and Village Industries Commission, various voluntary organizations and State level Federations. Few organized palm complexes are planned in Orissa recently. Summing up the entire potentiality in brief Table 3 and Table 4 gives potentiality of a single palmyra palm and single date palm tree respectively.

Table 11.3
Potentiality of a Single Palmyra Palm Tree

Name of raw materials	*Quantity available*	*Finished product*	*Quantity*
Neera	150 litres	Gur	24 kgs.
Fibre	1 kg	Basket	2 Nos.
Eark	2.5 kg	Brush	12 Nos.
Leaf	8 nos.	Mats	6 Nos
Naar	16 nos.	Basket	1 No.

Table 11.4
Potentiality of a Single Date Palm Tree

Nameof raw material	*Quantity*	*Finished product*	*Quantity*
Neera	120 litres	Gur	15 kg.
Leaf	15 kg.	Rope	130 ft.
Petiole	20 Nos.	-	-

In many regions neera is fermented into alcoholic beverage called toddy. Toddy is certainly a better alcoholic drink than the commercial distilled liquor normally available in the market. From this alcohol yeast can be separated and used in bakery industry. Except tapping of neera, entire palm industry discussed above is the domain of women.

Knitting is certainly the exclusive creative work of women. It is either done as creative work or as part time occupation or even as full time employment. Traditionally, it has developed either in European countries or colder region of Third World countries where woollen garments were produced at home by women out of wool. It has now gone to even hotter tropics where it has ensured income for women with knitted woollen garments exported to colder region or knitting with cotton or synthetics were done suiting to local environment. In any case it has come to stay in third world countries ensuring income for women. Small automatic knitting machines were developed during past few years which ensures higher income.

Embroidery again is another job which is considered monopoly of women not only in third world countries but also round the globe. It is true be it is colourful embroidered fabrics of Africa or light coloured embroidery of Europe. In West Bengal and Bangladesh a type of embroidery work is done by women at home on old torn clothes. It is commonly known as "Kantha" Kantha with complicated and elaborate design is known as "Nakshi Kantha". Each district of undivided Bengal had its own typical motif on the Kantha they produce. The designs of various Kantha is kept is a museum at Nabasan between Calcutta and Kharagpur near Ghoraghata Railway Station organized by a voluntary organization called Ananda Niketan. The designs are so unique and beautiful that it has been adopted recently as modern style by doing embroidery work on new cloth for using as dress materials or sarees. It has become new style in Indian sub-continent. Besides Kantha stitch design, women all around third world produce beautiful designs of embroidery work on various fabrics giving various dimensions of artistic motifs.

Applique work again another style of fabric design created by women all over the world. Traditional applique work produced by artisans of Orissa be it is Pipili near Puri or in Phulbani or Ganjam are unique. It has its own style and designs with contrast colours that could earn reputation all over the world. Applique work has become a style everywhere and modern designs has been adopted particularly on children wear. This work is also the monopoly of women.

Large paper mills, no doubt, meeting our demand of paper that is needed in every walk of life. But these paper mills are eating away our bamboo wealth, forest wealth faster than they are created causing environmental decay. Thus we are now heading towards the world of less rain, more carbon dioxide, hotter atmosphere and so on. Traditional hand-made paper units on the other hand gives us more durable and varieties of paper using various conventional and non-conventional raw materials. It consumes various cellulosic wastes otherwise considered as rubbish. The total development is eco-friendly in nature. This system of manufacturing paper was first developed in China many centuries before birth of Christ which started spreading elsewhere. It has continued to stay in Japan, India and elsewhere even in Europe. In India, it has stayed for last several centuries in Tawang of Arunachal Pradesh, Sangener near Jaipur in Rajasthan, besides Bijanhari, Kalingpong, Dosghara, Moinan and Bhagawangola of West Bengal besides elsewhere.

In fact author has seen a 500 year old hand written book on paper in a museum at Bishnupur in Bankura district of West Bengal. Since small is beautiful, being eco-friendly it is being encouraged in many of the European countries. In India it is being encouraged by Khadi and Village Industries Commission and an UNDP sponsored project is being implemented in India. Hand-Made Paper can be produced adopting every easy technology which can be adopted by women. In fact various beautiful paper conversion products can only be produced by women with their artistic imagination using tender fingers. One can see this development in Abhoy Ashram near Calcutta to get first hand knowledge. In fact document papers or those paper item need to be kept for long period has to be hand-made paper.

All products collected from forest except perhaps timber is considered as minor forest produce. They can be various herbs, edible or non-edible oil seeds, lac, various edible items, fire wood and so on depending on the forest and region. Collection and selling of minor forest produce is done by women throughout the third world where mechanization yet to destroy the system. These minor forest produce are important raw materials of various industries like medicine, tanning, food and so on. Till now at least in Third World countries, this system of collections trying to maintain ecological balance since overdrawing is not possible and ensures adequate income to women thus maintaining social balance. Mechanisation will lead overdrawing of natural wealth and shall throw women out of employment bringing both social and physical environmental degradation as noticed in various developed countries. Similarly, Cassava or topioca, sometime considered as minor forest produce but often cultivated is considered not only as important source of starchy food but also source of industrial starch and a popular food in India called Sago. It was a tradition in Kerala where Cassava based food like sago was produced at home by women and finally packed and sold. Modernisation and mechanization has not only created unemployment but wealth was accumulated with privileged few ensuring widening of gap between poorest and richest contradicting national goal and independence.

Last but not least, we should also note that modern scientific development cannot be totally ignored. Many of the modern gadgets such as electronic items like radio, television, electronic gas lighters or mechanical items like mixies, juicers, etc. can be produced by assembling various parts. They do not required physical strength or even drudgery is almost absent. For the sake of social balance, assembling should be done by women only since the job suits them. It is reported that productivity of women is much higher. In India, several women organizations are formed who produce such items by assembling. But, threat remain from monopoly houses who do not allow these units to grow particularly due to aggressive publicity and marketing strategies, even sabotaging the interest of small units if it tend to grow. Threat is, in fact, much higher now after liberalization since various multinationals with money and muscle power has already entered the arena.

There are many such stories in Third World countries which need to be studied and preserved for our healthy and sustainable future growth.

12

Neem: Its Potential in Pest and Vector Management

Ramesh C. Saxena

Abstract

Producing enough food for the growing population, protecting plant-, animal-, and human health, and at the same time conserving the environment would require pest management approaches which are effective, ecologically sound, equitable and ethical. Here the potential of neem, Azadirachta indica A. Juss. (Meliaceae), for eco-friendly, sustainable pest- and vector management is described. The scope of neem in fertiliser management and improving soil fertility also is mentioned. The environmental service rendered by neem and its other benefits to humankind are given. Finally, increased income generation and employment opportunities in rural areas through increased propagation, collection, and processing of neem, for production of natural pest control materials, medicinal preparations, pharmaceuticals, toiletries, and even fertility regulating agents for domestic and export markets, are discussed. In the coming millenium, tropical countries in Asia, Africa and Latin America can gain a lot from increased awareness of neem's potential.

Keywords: *Azadirachta indica,* neem, pest management, vector management, environmental conservation.

The Global Human population has already crossed the six billion marks. Another one billion people will inhabit the earth during the present decade. By 2020, the world population will be ~8 billion people. The bulk of this population increase will occur in Asia and Africa, where about 840 million people are food insecure. Providing adequate food entitlements, safeguarding public health, meeting fuel and firewood needs, and at the same time preventing deforestation and conserving the environment, and slowing down runaway population growth will be daunting challenges for the developing world in the coming millennium. According to a recent report by the Washington-based International Food Policy Research Institute, by 2020, the world will be an even unfair place than it is at present, with food surpluses in the industrialized world and with chronic instability and food shortages in the south, particularly in African countries. By 2050, the scenario may become worse for food importing countries as with even 1 per cent growth in population levels, countries such as USA may cease to become food-exporting countries.

In the past two decades, "green revolution technologies" have more than doubled the yield potential of rice and wheat, especially in Asia. These high-input production systems requiring massive quantities of fertilizers, pesticides, irrigation, and machines, however, disregard the ecological integrity of land, forests, and water resources, endanger the flora and fauna, and cannot be sustained over generations. Also, we cannot look to the sea in future as fishing stocks in many parts of the world are already in crisis due to over-fishing or pollution. To a great extent, future food security and economic independence of developing countries would depend on improving the productivity of biophysical resources through the application of sustainable production methods, by improving tolerance of crops to adverse environmental conditions, and by reducing crop and post-harvest losses caused by pests and diseases. Appropriate technologies, which do not assault the nature, would have key roles to play in ensuring food security, in improving public and animal health, and in rehabilitating and conserving the environment to safeguard the well being of the posterity. Instead of striving for more "green revolutions" with emphasis on miracle seeds, hard-hitting synthetic and engineered pesticides, and increased use of fertilizers, the future must look to natural ways and processes for augmenting agricultural productivity. In fact, all development efforts and activities, including pest management, should be within well-defined ecological rules rather than within narrow economic gains. Sustainable agricultural systems must be *efficient* and *ecologically sound* for long-term food sufficiency, *equitable* in providing social justice, and *ethical* in respecting both future generations and other species. For developing countries, the use of the neem tree, *Azadirachta indica* A. Juss., may provide a key component in more sustainable agricultural system, including pest and nutrient management, human health, and environmental conservation.

Neem–A Fascinating Tree

Neem, a member of the Meliaceae family, is a botanical cousin of mahogany. According to a report of an ad hoc panel of the Board on Science and Technology for International Development, *"this plant may usher in a new era in pest control, provide millions with inexpensive medicines, cut down the rate of human population growth and even reduce erosion, deforestation, and the excessive temperature of an overheated globe"* (National Research Council, 1992). Neem's other descriptions, such as *"nature's bitter boon"*, *"nature's gift to humankind," "the tree for many an occasion," the tree that purifies" "the wonder tree," "the tree of the 21st century,"* and *"a tree for solving global problems,"* are a recognition of its versatility. Its botanic name, *Azadirachta indica,* derived from Farsi, *"azad darakht i hindi"* literally means "the free or noble tree of the Indian subcontinent," suggesting that it is intrinsically free from insect and disease problems.

Neem is believed to have originated in Burma and the arid regions of the Indian subcontinent, where it has almost been semi-domesticated. During the last century, neem was introduced in arid zones of Africa. Today, it is grown in many Asian countries, in tropical regions of the New World, in several Caribbean and in some Mediterranean countries. During the last decade, neem has been introduced and planted on a large scale in Australia, in the Philippines, and in the Plains of Arafat near Mecca in Saudi Arabia, regions ecologically highly diverse. In all these areas, the tree is thriving – a testimony to its hardiness and adaptability. Neem, however, is unsuited to growing in cooler and mountainous areas (>1000 m).

Neem is an evergreen, tall, fast-growing tree, which can reach a height of 25 m and 2.5 m in girth (Fig. 1). It has an attractive crown of deep-green foliage (which can spread 10 m

Fig. 1. Neem tree–Nature's gift to mankind. The versatile tree is widespread in Asia and Africa, but lack of awareness of its potential in Africa led someone to fell a full-grown tree *(foreground),* which could have been a rich source of natural pest control materials and other useful products *(Photo by RC Saxena).*

across) and masses of honey-scented flowers. The tree thrives even on nutrient-poor dry soil. It tolerates high to very high temperatures, low rainfall, long spells of drought, and salinity. It is propagated by seed; 9 to 12 month-old seedlings transplant well. Birds and fruit bats also disperse the seed. Fruiting begins, in 3 to 5 years. In the Indian subcontinent, neem flowers from January through April and fruits mature from May to August. In coastal Kenya, fruiting occurs in March and April; some off-types also fruit in November or December. The fruit is about 2 cm long and, when ripe, has a yellow fleshy pericarp, a white hard shell, and a brown, oil-rich seed kernel. Fruit yields range from 30 to 100 kg per tree, depending on rainfall, insolation, soil type, and neem ecotype or genotype (Fig. 2). Fifty kg of fresh fruit yields 30 kg of seed, which gives about 6 kg of oil and 24 kg of seed cake. Seed viability ranges from 6 to 8 wk, but thoroughly cleaned and properly dried and cooled seeds remain viable up to 6 months. Propagation by stumps and stem cuttings also is practiced. Plantlets, produced by tissue culturing, have also been used for propagation with partial successes.

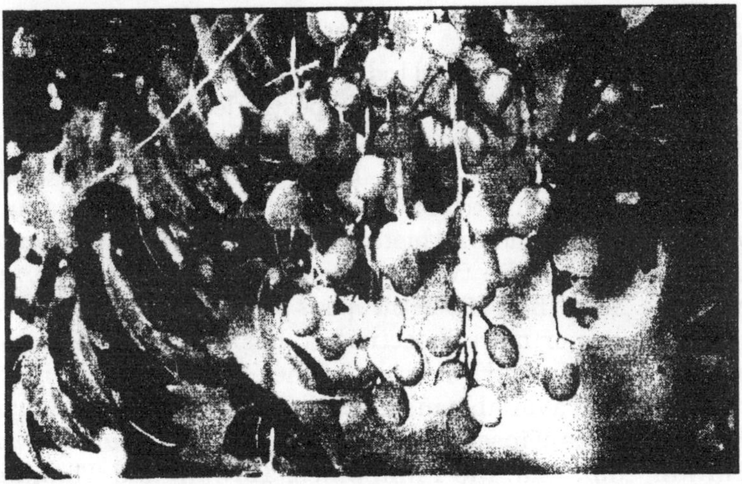

Fig. 2. A full-grown neem tree produces 50 to 80 kg of fruits; 50 kg of fruits yield 30 kg of seed from which 6 kg of highly valuable oil and 24 kg of cake can be recovered *(Photo by RC Saxena).*

Neem is bitter in taste. The bitterness is due to the presence of an array of complex compounds called "triterpenes" or more specifically, "limonoids." More than 100 bio-active compounds have been isolated from various parts of the neem tree; still more are being isolated. This formidable array of highly bio-active compounds makes neem a unique plant with potential applications in agriculture, animal care, public health, and for regulating even human fertility. The limonoids in neem belong to nine basic structure groups: *azadirone* (from oil), *amoorastaitin* (from fresh leaves), *vepinin* (from seed oil), *vilasinin* (from green leaves), *gedunin* (from seed oil and bark), *nimbin* (from leaves and seed), *nimbolin* (from kernel), and *salannin* (from fresh leaves and seed), and the *aza* group (from neem seed). *Azadirachtin* and its analogs have fascinated researchers for the past 30 years because of phagorepellency, growth inhibition, and chemosterilzing effects on insect pests (Saxena 1989, Schmutterer 1990, 1995). The azadirachtin content in neem could vary considerably due to edaphic, climatic, or genotypic differences.

Neem for Eco-friendly Pest and Vector Management

Crop Pests

Pest control as practiced today in most developing countries relies mainly on the use of imported pesticides. This dependence has to be reduced. Although pesticides are generally profitable on direct crop returns bases, their use often leads to the contamination of terrestrial and aquatic environments, damage to beneficial insects and wild biota, accidental poisoning of humans and livestock, and the twin problems of pest resistance and resurgence. More than 500 arthropods pest species have become resistant to one or more insecticides. Resistance of the cotton boll worm in India and Pakistan, of the Colorado potato beetle in the USA to all available insecticides, and of the diamondback moth to all classes of insecticides, including *Bacillus thuringiensis,* in Hawaii, Malaysia, the Philippines, Taiwan, and Thailand, illustrate the complexity of the problem. Shifts in pest status – from minor to major, and resurgence of pests, such as white flies, caused by direct or indirect destruction of pests' natural enemies are other unwelcome developments associated with pesticide use. A World Health Organization and United Nations Environment Programme report estimated that there are one million human pesticide poisonings each year in the world, with about 20,000 deaths, mostly in developing countries. The problem is rendered even more difficult because few, if any, new compounds are coming to replace old insecticides. The cost of developing and registering new pesticides is staggering: almost US$ 60 million, and pesticide manufacturers are unwilling to risk investments on products whose market life could be shortened by development of pest resistance.

For ecologically sound, equitable, and ethical pest management, there is a need for control agents that are pest-specific, non-toxic to humans and other biota, biodegradable, less prone to pest resistance and resurgence, and relatively less expensive. Among various options, *neem has been identified a source of environmentally "soft" natural pesticides.*

Neem has had a long history of use primarily against household and storage pests and to some extent against crop pests in the Indian subcontinent. As early as 1930, neem cake was applied to rice and sugarcane fields against stem borers and white ants. Early observations that swarming locusts did not attack neem leaves have been confirmed in laboratory studies and attributed to neem's antifeedant activity against locusts.

The pest control potential of neem in developing countries, however, remained largely untapped due to the advent of broad-spectrum synthetic insecticides. Also, publicity given to slogans such as "the only good bug is a dead bug" and identifying traditional uses of neem as backward, gradually weaned people away from using neem. It is only in the past decade that the pest control potential of neem has been appreciated. Though subtle, neem's effects such as repellence, feeding and oviposition deterrence, growth inhibition, mating disruption, chemo-sterilization, etc. are now considered far more desirable than a quick knock-down in integrated pest management programmes as they reduce the risk of exposing pests' natural enemies to poisoned food or starvation. In spite of high selectivity, *neem derivatives affect ca. 400 to 500 species of insect pests* belonging to different orders, one species of ostracod, several species of mites, and nematodes, and even noxious snails and fungi, including aflatoxin-producing *Aspergillus* spp.

Results of field trials in some major food crops in tropical countries will illustrate the value of neem-based pest management for enhancing agricultural productivity in Asia and Africa.

Rice. The efficacy of neem derivatives against major pests of rice and virus diseases transmitted by them and increases in yield has been reviewed by Saxena (1989). In the Philippines, application of a 2:10 neem cake urea mixture at 120 kg/ha reduced the incidence of ragged stunt, grassy stunt, and tungro viruses and significantly increased the rice yield more in both dry and wet seasons. Also, weekly spraying of 50 per cent neem oil-custard-apple oil mixture in 4:1 proportion (vol/vol) at 8 l/ha from seedling to the maximum tillering stage decreased the tungro incidence and increased the yield. The low input cost of the treatment contributed to a high net gain compared with the insecticide treatment. In India, neem treatments controlled populations of the green leafhopper, the yellow stem borer, the rice gall midge, and grasshoppers. Plots sprayed with 2 per cent neem seed extract at 10 kg/ha yielded the highest grain yield.

Fig. 3. A bountiful harvest was obtained when powdered neem cake was applied @ 1 g/plant into maize plant whorl at 4 weeks after plant emergence at 4 and 6 WE. The grain yield with neem was on par with that obtained with insecticides and almost double of untreated control *(Photo by RC Saxena).*

Maize and sorghum. In trials conducted at the ICIPE Field Station and in farmers' fields in Kenya, foliar application of powdered neem seed at 3 g/plant or powdered neem cake at 1 g/plant once at 4 weeks after crop emergence (WE) or twice at 4 and 6 WE to maize, which had been infested with the spotted stem borer, significantly reduced the foliar damage, stem tunneling, tassel breakage, and populations of borer larvae. Grain yield in neem-treated maize plots was as high as that obtained with insecticides and significantly higher than that in untreated control plots (Fig. 3, Table 1). Storage of neem cake up to 2 years in the dark did not reduce its pest effectiveness (Fig. 4, Table 2). Similar reduction in pest damage, including their body size as measured by the width of larval head capsules, and increase in yield were obtained when neem cake was applied to the sorghum crop (Table 3).

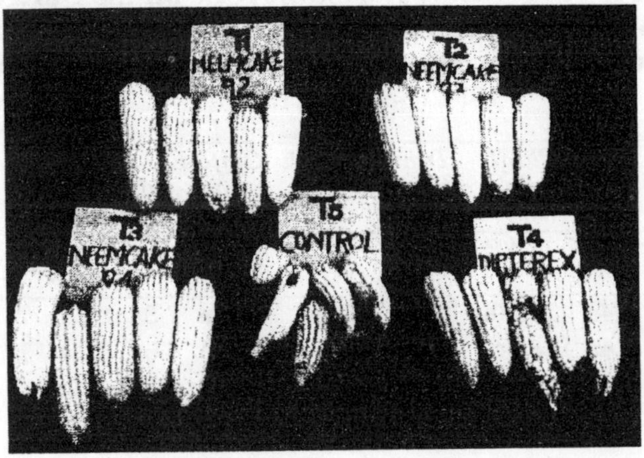

Fig. 4. Cobs harvested from neem-treated, insecticide-treated, and untreated maize fields. Compared with cobs in untreated or insecticide-treated plants, the grain quality was much superior in cobs harvested from pants treated with fresh, 1-year-old, or 2-year-old neem cake *(Photo by RC Saxena).*

Table 12.1
Tassel Breakage by *Chilo Partellus* Larvae and Grain Yield in Plots Planted to Stemborer-Susceptible 'Katumani' Maize Cultivar and Applied with Neem Seed Powder (NSP) or Furadan. ICIPE Field and Farmer's Field, Mbita, Short-rains Cropping Season 1992 (Saxena, Unpubl.)[1]

Treatment	*ICIPE Field Station*		*Farmer's Field*	
	Tassel breakage (%)	*Yield (kg/ha)*	*Tassel breakage (%)*	*Yield (kg/ha)*
NSP (basal)	17.0b	4530b	12.0b	3570b
NSP (foliar)	2.0a	6430a	4.0a	5480a
NSP (basal+ foliar)	2.0a	5870a	2.0a	5630a
Furadan 5G	0.3a	6400a	0.3a	6130a
Untreated (control)	21.0b	3370b	17.0c	3850b

[1]Within a column, means followed by a common letter are not significantly different at the 5 per cent level by the LSD test; averages of 4 replications.

Table 12.2
Infestation and Plant Damage by *Chilo partellus* and *Eldana Saccharina* Larvae and Grain Yield in Plots Planted to Moderately Resistant ICZ5 Maize Cultivar and Applied with Neem Cake (NC) or an Insecticide. ICIPE Field Station, Mbita, Long-rains Cropping Season 1994 (Saxena, 1998)[1]

	Damage[2]	*Plant height (cm)*	*Tunnel length (cm)*		*Chilo*	*Eldana*	*Per cent tassel*	*Yield*
Treatment	*9WE*	*14WE*	*7WE*	*14WE*	*(no.)[3]*	*(no.)[4]*	*Breakage[5]*	*(kg/ha)*
Fresh NC	2.4a	180ab	0.6a	15.6a	1.5a	7.8a	8.6a	7458a
1-yr-old NC	2.4a	186a	0.7a	13.0a	1.5a	6.5a	8.1a	7760a
2-yr-old NC	2.4a	185a	0.6a	18.5a	0.3a	9.4a	9.4a	7469a
Dipterex	2.1a	177ab	0.5a	20.3ab	1.5a	17.0a	17.0a	7271a
Untreated (control)	4.6b	167b	7.8b	31.6b	14.3b	25.5b	35.5b	5088b

[1] Within a column, means followed by a common letter are not significantly different at the 5 per cent level by LSD test; averages of 4 replications. [2] Foliar damage scored visually on 1-9 scale (1= no damage, 9 = completely damaged).

[3,4,5] *Chilo* larvae were recorded at 9 WE; *Eldana* larvae at 11 WE, and tassel breakage at 11 WE.

Banana: The banana weevil, *Cosmpolites sordidus,* and parasitic nematodes are major pests of banana and plantain. They often occur together in the same plant and may destroy the cor and the root system, resulting in loss of fruit yield. Most of the highland bananas in Eastern Africa are highly susceptible to the weevil and nematode infestations. Soil applications of NSP or NC at 100 g/plant at planting and, subsequently, at 3-month intervals, reduced the populations of the root-lesion nematode, *Pratylenchus goodeyi* and the root-knot nematodes, *Meloidogyne* spp., on par with Furadan 5G applied at 40 g/plant at planting and then at 6-mo-intervals to banana plants grown in 100 containers with controlled levels of banana nematode infestations (Musabyimana and Saxena 1999a). Eight months after planting, banana plants treated with NC, NSP, kernel

Table 12.3

Infestation and Plant Damage by *Chilo Partellus* Larvae and Grain Yield in Plots Planted to Stemborer-Susceptible 'Serena' Sorghum Cultivar and Applied with Neem Cake (NC) Once at 4 Weeks After Crop Emergence (WE) or Twice at 4 and 6 WE, or with Dipterex. Mbita, Short-Rains Cropping Season, 1994 (Saxena 1998)[1]

	Experiment Station						*Farmer's Field*					
Treatment	*Dam-age 9WE*	*Plant height (cm) 9WE*	*Tunnel length (cm) 15WE*	*Larvae (no.) 9WE*	*Head width (mm)*	*Yield (kg/ ha)*	*Dam a-ge 9WE*	*Plant height (cm) 9WE*	*Tunnel length (cm) 15WE*	*Larvae (no.) 9WE*	*Head width (mm)*	*Yield (kg/ha)*
NC once	2.9ab	110a	27.5ab	7.4ab	0.66a	6182ab	2.9a	114a	25.4b	18.8ab	0.64a	5242a
NC twice	2.9ab	106ab	21.3a	2.2a	0.67a	7312a	3.0a	112a	18.9a	16.8a	0.63a	5052a
Dipterex	2.3ab	117a	20.3a	7.0ab	0.79ab	6523ab	2.5a	111a	22.9ab	19.4ab	0.96c	5043a
Untreated (control)	3.5b	94b	30.3b	10.2b	0.84b	6056b	4.1b	99b	27.9b	25.8b	0.90b	3908b

Within a column, means followed by the same letter are not significantly different at the 5 per cent level by the LSD test; averages of 4 replications.[2] Foliar damage scored visually on 1 to 9 scale (1 = no damage; 9 = completely damaged).

powder, or with oil had 7 to 95 times less parasitic nematodes than the untreated control. However, only NC or NSP applied to unpared banana plants kept the nematode population below the economic threshold (Table 4). At 8 months after incorporation into the soil, NC or NSP application was still effective against banana nematodes, while the nematicidal activity of Furadan seemed to decline. Weevil larvae fed little on or avoided altogether neem-treated corms, while extensive feeding and damage occurred on untreated corms. With neem treatment, fruit yield increased by 27-50 per cent over the control in the first crop and by 30-60 per cent over the control in the second crop (Fig. 4), but yield with Furadan during the second crop was even less than that in the control (Musabyimana and Saxena 1999b).

Table 12.4

Effect of Soil Application of Neem Seed Powder (NSP), Neem Cake (NC), Neem Kernel Powder (NKP), or Treatment with Neem Oil (NO) on Population of Banana Nematodes at 2 and 8 Months after Treatment of Pared or Unpared Suckers Planted in Drums. Mbita Point Field Station (Musabyimana and Saxena 1999a)[1]

	Nematode Population at (no. / 100g of roots)± SEM	
Treatment	*2 months*	*8 months*
Pared	1200± 489a	22200+3747a
Unpared + NSP	300± 300a	3600± 490a
Pared + Furadan	0+0a	16800±2135a
Pared + NC	0+0a	12000+2135a
Pared+ NSP	0+0a	22500±2265a
Pared + NKP	0+0a	81600+23510b
Unpared + NC	300+300a	1200+0a
Unpared + NO	0+0a	5700+1025a
Unpared + NKP	125+125a	27600+3730a
Unpared (untreated)	25050+4057b	114000+4673b
CV per cent	95.7	50.4
Difference	**	**

[1] Means in columns followed by the same letter do not differ significantly (P < 0.05; Tukey's test); avg. of 4 replicates; ** = JP < 0.01 (Tukey's test).

Grain legumes and vegetables. Because of high profitability of legumes and vegetables, farmers tend to overuse chemicals, which results in hazard to the environment and health of producer and consumer, as well as serious resistance problems. However, neem can provide satisfactory control of insect pests affecting grain legumes and vegetables.

Fig. 4. Difference in size of 'Nakyetengu' banana Bunches harvested from an untreated plot *(left)* and from neem-treated plot *(right)* in a farmer's field in Oyugis, Western Kenya, 1998 *(Photo by T. Musabyimana).*

A wide range of pests attacks cowpea, a major food crop in Africa. Sprays with neem seed extracts (NSE) were quite effective against lepidopterous pests, but weekly ULV spray applications of neem oil did not control the flower thrips, *Megalurothrips sjostedti* (Dreyer 1987). On the other hand, in trials conducted at ICIPE's Experiment Station and in a farmer's field in Kenya, applications of 2 or 3 per cent (NSE) at 200 lha at 38, 47 and 51 days after emergence (DE) of cowpea crop or ULV spray application of 5, 10, or 20 per cent NSE at 10 lha significantly reduced the number of thrips larvae in flowers recorded 2 days after each treatment (Saxena and Kidiavai 1997). Also fewer adults occurred in flowers at 51 DE in plots sprayed with 5 per cent, 10 per cent, or 20 per cent NSE. Grain yield was significantly higher in plots sprayed with 20 per cent NSE than in untreated control plots and was comparable to yield obtained in plots sprayed thrice with cypermethrin (Table 5). Because of the low cost of NSE treatment, the net gain was often more when cowpea was sprayed with NSE than with the insecticide. Also, grain quality was superior in neem-treated plots than in control or cypermethrin-treated plots (Fig. 5).

In common beans, high volume spray applications of 2 per cent neem kernel extract (NKE) at 11 day intervals effectively controlled the chrysomelid beetle, *Ootheca bennigseni* (Karel 1989).

Kales, cabbages, cauliflowers, and other crucifers are highly vulnerable to attack by the diamondback moth (DBM), *Plutella xylostella*. Synthetic insecticides do not provide long-term control as the pest is known to develop resistance rapidly (Talekar and Shelton 1993). In trials conducted in Togo, weekly high volume (HV) spray applications of a 4 per cent methanolic NKE (Adhikary 1985) or even 2.25 to 5 per cent aqueous NKE (Dreyer and Hellpap 1991) almost completely protected the cabbage. In repeated trials conducted at ICIPE's Experiment Station and in a former's field in Western Kenya, ULV spray applications of 20 per cent neem seed extract (NSE) at 10-d-intervals, significantly reduced the pest infestation and damage and increased the yield of marketable leaves of kale ("Sukuma wiki" in Kiswahili) (Table 6, Fig. 6). Spray applications of 0.3 per cent neem extractive were not as good as NSE treatment. The population of spiders, which are important predators of DBM larvae, was as high in neem-treated plots as in untreated control plots, while it was much lower in cypermethrin-treated plots. During long rainy season, kale leaf yield was significantly higher with NSE than with cypermethrin. Economic returns with NSE treatments were promising, as the cost of neem seed is low.

Table 12.5
Comparative Yield and Value of Cowpea Grain after Deducting the Cost of Neem Seed Extract (NSE) or Cypermethrin Applied Thrice to Cowpea Crop. Mbita Point. Long-rains Cropping Season, 1994 (Saxena and Kidiavai 1997)[1]

Treatment[2]	*Grain yield (kg/ha)*	*Value of yield ($/ha)*	*Cost of treatment ($/ha)*	*Value of yield minus cost of treatment ($/ha)*
		Field Station		
NSE 5%	1160ab	580	1.5	578.5
NSE 10%	1280ab	640	3.0	637.0
NSE 20%	1480a	740	6.0	734.0
Cypermethrin	1480a	740	108.0	632.0
Control (1 per cent Teepol)	1050b	525	0.0	525.0
		Farmer's Field		
NSE 5%	1450d	725	1.5	723.5
NSE 10%	1630c	815	3.0	812.0
NSE 20%	1760b	880	6.0	874.0
Cypermethrin	2130a	1065	108.0	957.0
Control (1% Teepol)	1290e	645	0.0	645.0

[1]Means followed by a common letter do not differ significantly (*P* < 0.05; Ryan-Einot-Gabriel-Welsch Multiple Range Test). [2]Using ULV applicators, NSE or Cypermethrin was applied 3 times at 10 1/ha at 31, 39, and 49 DE. [3]US$ = KSh 54; cost of cowpea grain = US$ 0.50/kg. [4]Treatment cost includes only cost of NSE or Cypermethrin; neem seed @ US$ 0.5/kg, Cypermethrin @ US$ 36/l.

Fig. 5. *(clockwise)* Cowpea pods harvested from plots sprayed with Cypermethrin, Teepol (control), or with 20 per cent, 5 per cent, or 10 per cent neem seed extract (NSE). Pods harvested from plots sprayed with 20 per cent NSE have a reddish pigmentation with uniformly ripened grains, which had an appealing sheen; pods from Cypermethrin-treated plots had grains, which did not ripen evenly *(Photo by R. C. Saxena).*

Fig. 6. Negligible pest damage and luxuriant foliage in NSE-sprayed kale plants *(left)* and heavy damage and sparse foliage in untreated kale plots. ICIPE Field Station, Long-rains Cropping Season 1998 *(Photos by RC. Saxena).*

1. Means followed by a common letter do not differ significantly ($P < 0.001$ for Short rainy season 1997 and for Site # 1 Long Rainy Season 1998; $P < 0.002$ for Site # 2 Long-rains Cropping Season 1998; Ryan-Einot-Gabriel-Welsch Multiple Range Test); avg. of 5 replications.
2. Using ULV applicators, NSE, NE or Cypermethrin was applied to kale crop at 10 lha at 10-d-intervals.
3. A natural antioxidant used as a spice in India.

In Africa, the root-knot nematodes, *Meloidogyne* spp., and the fruit borer, *Helicoverpa armigera,* and are the most damaging pests of tomato. As nematodes are "unseen enemies," their role in limiting tomato production in tropical regions is generally overlooked. Rossner and Zebitz (1987) and Parveen and Alam (1998) reported nematicidal effects of neem materials in tomato. In field trials conducted at ICIPE's Experiment Station in 1997-98, we found that, compared with untreated control, soil application of NSP at 3 g/hill at planting significantly reduced the number of galls per plant on par with Furadan (Table 7, Fig. 7). However, the nematicidal effect of NSP treatment persisted longer than that of Furadan. The fruit borer incidence was low, but as a precautionary measure against insect pests in general, NSE, NE, or cypermethrin was sprayed with a ULV applicator at 10 l/ha at 10-d intervals. Although fruit yield did not increase significantly with neem or pesticide treatment over the untreated control, the quality of fruits produced in neem-treated plots was distinctly superior (Fig. 7). In Niger, weekly spray applications of 5 per cent aqueous NSE reduced the tomato fruit borer damage and increased the marketable fruit yield (Ostermann 1992).

Okra pests, such as the leaf-eating caterpillar, *Syllepta derogata,* was quite susceptible to spray applications of even at 0.25 per cent aqueous NKE (Dreyer 1987) or 5, 10, or 20 per cent aqueous NSE (Cobbinah and Olei-Owusu 1988). Also, the cotton aphid, *Aphis gossypii,* was well-controlled on okra by four weekly sprays of 0.5 per cent aqueous NSE or 2 per cent neem oil; the effects being on par with butocarboxim insecticide (Dreyer and Hellpap 1991).

Table 12.6

Comparative Diamondback Moth (DBM), *Plutella Xylostella*, Infestation, Spider Population, and Leaf Yield in Fields Planted to Highly Susceptible 'Southern Georgia' Kale Cultivar and Sprayed with Neem Seed Extract (NSE), Azadirachtin-Rich Neem Extractive (NE), or with Cypermethrin. ICIPE Field Station, Mbita (Saxena, Unpubl.),

Treatment[2]	*DBM (no.)/20 leaves*	*Spiders (no.)/20 leaves*	*Leaf yield (kg/ha)*
	Short-rains Cropping Season 1997		
NSE 20%	41.8 ± 6.89a	-	14043 ± 1577a
NSE 20% ± ajwan[3]	51.0 ± 9.98a	-	13737 ± 1497a
Cypermethrin	45.2 ± 7.21a	-	13972 ± 1931a
Untreated (control)	109.2 ± 14.86b	-	6630 ± 741b
	Site # 1. Long-rains Cropping Season 1998		
NSE 20%	24.6 ± 3.82a	15.2 ± 3.79a	9841± 919.2a
NE 0.3%	51.6 ± 7.22b	16.0 ± 7.22a	6187 ± 482.1b
Cypermethrin	50.8± 6.04b	5.6 ± 1.21a	6131 ± 854.0b
Untreated (control)	85.8 ± 11.19c	17.6 ± 2.84a	5360 ± 947.4b
	Site # 2. Long-rains Cropping Season 1998		
NSE 20%	10.2 ± 1.74a	18.0 ± 2.81a	20776 ± 1789a
NE 0.3%	14.6 ± 1.33a	21.6 ± 3.97a	12591 ± 1111b
Cypermethrin	16.2 ± 2.84a	8.6 ± 1.03a	14117 ± 1522b
Untreated (control)	35.6 ± 4.32b	20.2 ± 2.69a	13979 ± 1108b

Table 12.7
Comparison of Root Galls Caused by Root-Knot Nematodes in Neem-treated, Furadan-Treated, and Untreated Tomato Plants. ICIPE Field Station, Mbita (Saxena, Unpubl.),

Treatment[3]	*Root gall score (0-10 scale)*[2] *recorded*	*Recorded at weeks after transplanting*	
	4WT	*6WT*	*8WT*
NSP (3 g/plant), NSE 20 per cent	2.1± 0.30a	1.4 ± 0.18a	
NSP (3 g/plant), NE 0.3 per cent	1.6 ± 0.47a	1.3 ± 0.26a	
Furadan 5G (1 g/plant), Cyper.	2.8 ± 0.33ab	2.4 ± 0.28a	
Untreated (control)	3.9 ± 0.49b	2.2 ± 0.32a	
	Short-rains Cropping Season 1998		
NSP (3 g/plant), NSE 20 per cent	1.8 ± 0.50a	0.9 ± 0.17a	
NSP (3 g/plant), NE 0.5 per cent	2.1 ± 0.66a	2.2 ± 0.25b	
NSP (3 g/plant), NE 1.0 per cent	2.0 ± 0.44a	1.1 ± 0.10a	
Furadan 5G (1 g/plant), Cyper.	2.2 ± 0.45a	0.9 ± 0.17a	
Untreated (control)	4.4 ± 0.35b	4.7 ± 0.39c	
	Long-rains Cropping Season 1999		
NSP (3 g/plant), NSE 20 per cent	1.6 ± 0.43a	1.4 ± 0.39a	0.8 ± 0.34a
NSP (3 g/plant), NE 10 per cent	1.9 ± 0.51a	1.7 ± 0.34a	1.5 ± 0.35ab
Furadan 5G (1 g/plant), Cyper.	1.9 ± 0.28a	2.0 ± 0.23a	2.0 ± 0.36bc
Untreated (control)	3.1 ± 0.22b	3.5 ± 0.31b	2.8 ± 0.17c

1. Means followed by a common letter do not differ significantly *(P* < 0.001-0.05 for different cropping seasons; Ryan-Einot-Gabriel-Welsch Multiple Range Test); avg. of 5 replications for 1997 and 1999; avg. of 4 replications for 1998.
2. Root gall score (on 0-10 scale) : 0 = 0 galls per plant, 1 = 1-2,2 = 3-5, 3 = 6-10,4 = 11-15, 5 = 16-20, 7 = 31-40, 8 = 41-50, 9 = 51-60, 10 = 61 and above galls per plant.

3. Using ULV applicators, NSE, NE or Cypermethrin was applied to tomato crop at 10 l/ha at 10-d-intervals.

In Niger, foliar applications of aqueous NSE 0.25 per cent, 0.5 per cent or 1 per cent to amaranth fields strongly repelled *Spodoptera exigua,* while a soil drench of 0.5 per cent NSE repelled *Spodoptera littoralis* (Ostermann 1992). Spray applications at 0.5 per cent or 1 per cent NSE reduced the foliar damage by *S. exigua* significantly, while pre- and post-sowing soil drenches with 0.5 per cent NSE at 1000 l/ha stopped the immigration of *S. littoralis* larvae into treated fields and almost doubled the leaf yield over that of the untreated plots (Ostermann 1992).

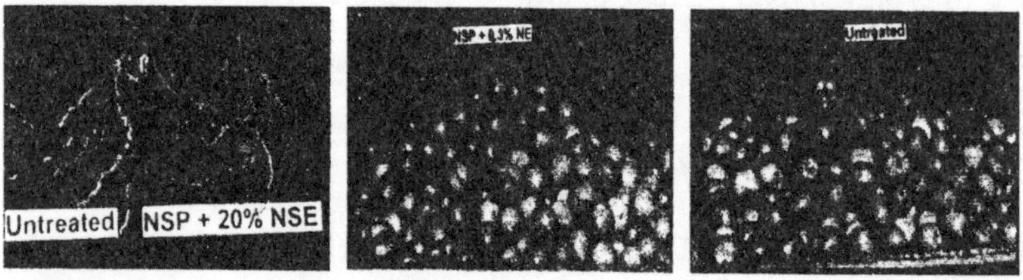

Fig. 7. Galled roots and healthy roots from untreated and neem treated tomato plants *(left);* difference in quality of fruits harvested from neem-treated plots *(middle)* and untreated plots *(right) (Photos by R.C. Saxena).*

In Sudan, remarkable results were obtained with neem products in the control of the sweet potato whitefly, *Bemisia tabaci,* and the leafhopper, *Jacobiasca lybica* on potao (Siddig 1987, 1991). Two HV applications of 2.5 per cent aqueous NKE sprayed at fortnightly intervals significantly reduced the pest populations to >50 per cent of the control and increased the yield. The potato tuber moth *Phthorimaea operculella,* was unaffected in the field but spray applications of 0.05 per cent and 0.1 per cent of neem oil strongly deterred oviposition and prevented damage in the stored potatoes (Siddig 1988).

Agroforestry and tobacco. Insects and nematodes also affect trees and crops in agroforestry. In collaborative trials conducted by International Centre for Research in Agroforestry (ICRAF) at Shinyanga in Tanzania in 1995-1996, application of powdered NC at 2 g/plant to a hybrid maize, "Cargill," at 4 and 6 weeks after sowing, registered a 30 per cent yield increase over the untreated control (ICIPE 1998). Application of NC at 135 kg/ha also reduced the termite damage and significantly increased the grain yield of hybrid maize over the Furadan-treated or untreated crop. In a long-term trial conducted at ICRAF field Station at Machakos, it was observed that when NC was applied at 15 g/grevillea seedling, the tree mortality after 15 months was 60 per cent, compared with 52 per cent tree mortality in Furadan-treatment, and 72 per cent in untreated control.

In field trials conducted in Tabora, Tanzania, although application of NSP or NC at 15 g/m^2 was not as effective as ethylene dibromide at 62 ml/m^2 in reducing the root galling index in tobacco plants, the tobacco yield increased significantly with neem treatments (ICIPE 1998).

Pests of Stored Products

Post-harvest losses are notoriously high in developing countries. Worldwide annual losses in store reach up to 10 per cent of all stored grain, i.e., 13 million tons of grain lost due to insects or 100 million tons to failure to store properly. Saxena (1995) has recently reviewed the potential of neem against pests of stored products: grain legumes, maize, sorghum, wheat, rice and paddy, and potato tubers. At farm level storage and warehouses, the application of neem derivatives to bags and stored grains has provided protection against insect pests. Powdered neem seed kernel mixed with paddy (1 to 2 per cent) significantly reduced infestation in warehouses. Neem leaves mixed with paddy (2 per cent), bags treated with 2 per cent neem extract, or 20 to 30 cm dried neem leaf barrier between the bags and storage floor significantly reduced insect infestation and damage to grain during a 3 month storage period; the effectiveness being comparable to methacrifos dust. Likewise, neem seed extract at 7.2 g/90 kg capacity jute bag (100 × 60 cm) controlled 80 per cent of the population of major insects and checked the damage to wheat up to 6 months. The treatment was effective up to 13 months and provided more than 70 per cent protection as compared with untreated control. The neem seed extract treatment was as effective as that of 0.0005 per cent primiphos methyl mixed with the grain. Using this technology in Sind, Pakistan, high benefit-cost ratios were obtained by small, medium, and large-scale farmers.

The effectiveness of neem oil alone or in combination with fumigation was evaluated against five major species of stored grain pests infesting rice and paddy grains in a warehouse trials conducted in the Philippines. Rice grain treated with 0.05 to 0.1 per cent neem oil or treated with neem oil after fumigation with 'Phostoxin,' and stored for 8 months had significantly less red flour beetle adults than in untreated control. Both kinds of neem treatments were as effective as the bag treatment with 'Actellic' at 25 ug/cm^2 or grain treatment with Actellic at 0.0005 per cent, and suppressed the pest population by 60 per cent. The population build-up also was reduced when either fumigated or non-fumigated rice was stored in bags treated with neem oil at >1 mg/cm^2. The lesser grain borer, the rice weevil, the saw-toothed grain beetle, and the rice moth were similarly affected by neem treatments alone or in combination with prior grain fumigation. Fumigation and Phostoxin were effective only for about 2 months against the lesser grain borer, and for up to 6 months against other pest species, while neem oil treatments were effective up to 8 months. Compared with the pest damage to untreated or fumigated rice, neem oil treatment significantly reduced the damage to rice grain. At 8 months after storage, weevil attacked grains in neem treatments were 50 per cent of those in the fumigated rice and 25 per cent of those in the untreated rice. Neem treatments also reduced the pest populations and the damage in paddy.

In studies conducted in Kenya, the growth and development of 1st instars of the maize weevil was completely arrested in maize grain treated with neem oil at 0.02 per cent, while the weight loss of treated cobs was less than 1 per cent as compared with a 50 per cent reduction in weight of untreated cobs stored for 6 months (Kega and Saxena 1996).

While neem treatments cannot replace completely chemical pesticides used in stored product preservation, the amounts of pesticides needed could be reduced, thereby decreasing the pesticide load in foodgrains. With proper timing and innovative methods of application, their use could be integrated in stored products management.

Blood-sucking Pests

The effects of neem on hematophagous insects affecting humans and livestock have been reviewed (Ascher and Meisner 1989). Application of a paste made from neem leaves and turmeric in 4:1 proportion to the skin cured 97 per cent of the patients suffering from scabies caused by the itch mite in 3-15 d. Monthly sprays of ethanolic extracts of neem or weekly bathing in azadirachtin-rich aqueous 1:20 'Green Gold' controlled the bush tick and the cattle tick in Australia, but were less effective against the brown dog tick (Rice 1993). In Jamaica, neem kernel extract controlled ticks on cattle and dogs.

Neem products repel and affect the development of mosquitoes. Two per cent neem oil mixed in coconut oil, when applied to exposed body parts of human volunteers, provided complete protection for 12 h from bites of all anophelines (Sharma et al. 1993). Kerosene lamps containing 0.01-1 per cent neem oil, lighted in rooms containing human volunteers, reduced mosquito biting activity as well as catches of mosquitoes resting on walls in the rooms; protection was greater against *Anopheles* than against *Culex*. Effectiveness of mats with neem oil against mosquitoes has also been demonstrated; the vapourizing oil repelled mosquitoes for 5-7 h at almost negligible cost. The sandfly also was totally repelled by neem oil, mixed with coconut or mustard oil, throughout the night under field conditions in India. Application of neem cake at 500 kg/ha, either alone or mixed with urea, in paddy fields in southern India reduced the number of pupae of *Culex tritaeniorhynchus,* the vector of Japanese encephalitis, and also resulted in higher grain yield.

Pest Resistance to Neem Materials

A few herbivorous insects, including some sucking insects, some beetles, and some moths do survive on neem but, largely, the tree is free from serious pest problems. Some insects can adapt to limonoids, but in laboratory tests two genetically different strains of the diamondback moth treated with a neem seed extract showed no sign of resistance in feeding and fecundity tests up to 35 generation. In contrast, deltamethrin-treated lines developed resistance factor of 20 in one line and 35 in the other. There was no cross-resistance between deltamethrin and neem seed extract in the deltamethrin-resistant lines. The diversity of neem compounds and their combined effects on insect pests seem to confer a built-in resistance prevention mechanism in neem. *However, wisdom demands that users should refrain from exclusive and extended application of single bio-active materials, such as azadirachtin.*

Neem for Soil Fertility and Fertilizer Management

Good soil fertility means good crop yields. Preventing the loss of plant nutrients from an ecosystem is important for soil-fertility management. N, P, K elements in soil in a 4:2:1 ratio determines the soil fertility; aberrations affect fertility and crop yield. Urea (with 46 per cent N) is generally applied to crops; but less than half of it in the form of nitrate is taken up by crops. The rest is lost through 'leaching,' or by *volatilization* or by *surface run-off.* Nitrate losses of 50 to 70 per cent through leaching can occur in rice crops. Ammonia volatilization in the form of gaseous due to denitrifying bacteria in soil can cause up to 60 per cent nitrate loss. These losses can be reduced either by split applications or use of 'slow-release urea,' or by adding nitrification inhibitors. Neem seed kernel extract has been found to be an excellent nitrification inhibitor, even better than sulfur-coated urea. Ammonia volatilization, urea hydrolysis, and leaching were all reduced when urea was blended or coated with neem cake. Neem cake coating of prilled urea increased nitrogen uptake by 4.5 to 19.4 per cent and increased rice and sugarcane yields correspondingly. Ready-to-use, neem-based urea-coating agents, such as 'Nimin' (containing ca. 5 per cent neem bitter tetranortriterpenoids) are now commercially available in India (Vyas et al. 1996). The bitters in neem delay the denitrification process up to 30 d by either killing nitrifying bacteria or suppressing their activity and effect a saving of up to 20 per cent of urea. The technology is becoming popular with farmers in India.

Environmental Service and Other Benefits

Environmental Service

Neem in Indian culture has been ranked higher than 'Kalpavriksha', the mythological wish-fulfilling tree. Although scientific studies are wanting, neem is reputed to purify air and the environment of noxious elements. Its shade not only cools but also prevents the occurrence of many diseases. During hot summer months in northern parts of the Indian subcontinent, the temperature under the neem tree is ~10°C less than the surrounding temperature. Restoration of the health of degraded soils and the ultimate use of such reclaimed wastelands lands through neem is another example of its value to humans. About a decade ago, some 50,000 neem trees were planted over 10 km^2 on the Plains of Arafat to provide shade for Muslim pilgrims during *hajj*. The neem plantation has had a marked impact on the area's microclimate, microflora, microfauna, and soil properties, and when full-grown could provide shade to 2 million pilgrims. It is believed that neem growing inside the house can help cleanse the surrounding air. Also, hanging neem twigs on the door of a house is said to reduce pollution.

The tree is not only beautiful to look at, providing grandeur and serenity, but also serves as a refugia to many beneficial organisms, bats, birds, honey bees, spiders, etc. Honeycombs established on the neem tree are free from the galleria wax moth infestation. Many species of birds and fruit-eating bats subsist on the sweet flesh of ripe fruits, while certain rodents selectively feed on the kernel, confirming neem's safety to warm-blooded animals. The litter of falling leaves improves soil fertility and the organic content. Recently, mycorrhizal associations between neem and bacterial and fungal endophytes have been identified. Indeed, the neem tree is a living microcosm!

The evergreen, perennial tree can survive from 200 to 300 years. Even a highly conservative estimate of the intangible 'environmental service' rendered by the tree at US$ 10 per month over its lifetime would give an astonishing value of US $ 24,000 to 36,000. Other tangible economic uses of neem and the benefits derived, such as biomass production, timber, seed, and honey, are quantifiable.

Reforestation and Agroforestry

Neem is a very valuable forestry species in India and Africa and also becoming popular in Tropical America, the Middle East countries, and in Australia. Being a hardy, multipurpose tree, it is ideal for reforestation programmes and for rehabilitating degraded, semiarid and arid lands, and coastal areas. During a severe drought in Tamil Nadu, India in June-July 1987, it was witnessed that neem grew luxuriantly, while other vegetation dried up.

Neem is useful as windbreaks. In areas of low rainfall and high wind speed, the neem tree can protect crops from desiccation. In the Majjia Valley in Niger, over 500 km of windbreaks comprised of double rows of neem trees have been planted to protect millet crops which reportedly resulted in a 20 per cent increase in grain yield. Neem windbreaks on a smaller scale have also been grown along sisal plantations in coastal Kenya. Large-scale planting of neem has been initiated in the Kwimba Afforestation Scheme in Tanzania and at Adjumani in northern Uganda.

In countries from Somalia to Mauritania, neem has been used for halting the spread of the Sahara desert. Also, neem is a preferred tree along avenues, in markets, and near homesteads because of the shade it provides. However, neem is best-planted in mixed stands. It was probably no coincidence that Emperor Ashoka, the great ruler of ancient India, in the 3rd century BC, commanded that the neem be planted along the royal highway and roads along with other perennials – tamarind and mahua. Neem has all the good characters for various social forestry programmes.

Neem is an excellent tree for silvipastoral system involving production of forage grasses and legumes. But according to some reports, neem cannot be grown among agricultural crops due to its aggressive habit. Others say that neem can be planted in combination with fruit cultures and crops such as sesame, cotton, hemp, peanuts, beans, sorghum, cassava, etc., particularly when neem trees are still young. The neem tree can be lopped to reduce shading and to provide fodder and mulch. Recent advances in tissue culture and biotechnology should make it possible to select neem phenotypes with desirable height and stature for use in intercropping and various agroforestry systems. The allelopathic effects of neem on crops, if any, need to be investigated.

Biomass Production and Utilization

Full-grown neem trees yield between 10 to 100 tons of dried biomass/ha, depending on rainfall, site characteristics, spacing, ecotype or genotype. Leaves comprise about 50 per cent of the biomass; fruits and wood constitute one-quarter each. Improved management of neem stands can yield harvests of about 12.5 cubic meter (40 tons) of high quality solid wood/ha.

Neem wood is hard and relatively heavy and is generally used for making carts, tool handles, farm implements, and even toys and religious icons in some parts of India. The wood seasons well, except for end splitting. Being durable and termite resistant, neem wood is used in making fence posts, poles for house construction, furniture, etc. There is growing market in some European countries for light-coloured neem wood for making household furniture. Polewood is especially important in developing countries; the tree's ability to resprout after cutting and to regrow its canopy after pollarding makes it highly suited to pole production. Neem grows fast and is a good source of firewood and fuels; the charcoal has high calorific value.

Neem in Human Health and Well-being and Animal Care

Medicinal Preparations, Pharmaceuticals, and Toiletries

Since ancient times, neem has been associated with healing. The early Ayurvedic literature, especially 'Charak Samhita,' and the Unani system of medicine document neem's medicinal and therapeutic uses. Even in Africa, Tropical America, and other parts of the world, where neem has been introduced recently, people have learnt of neem as herbal treatments for various ailments. A large number of medicinal, cosmetics, toiletries, and pharmaceuticals are now based on neem products.

Neem fruits, seed oil, leaves, bark, and roots have varied uses as general antiseptics, antimicrobials, treatment of urinary disorders, diarrhoea, fever, bronchitis, skin infections (Fig. 8), septic sores, infected burns, hypertension, and inflammatory diseases. Neem leaf extracts and teas are reputed to treat malaria, the antimalarial action is attributed to gedunin, a limonoid. Neem tablets, containing a neem leaf decoction, are recommended and are available in India for cases of diabetes mellitus of adults, nonketonic diabetes, and for cases of insulin fast or insulin sensitivity.

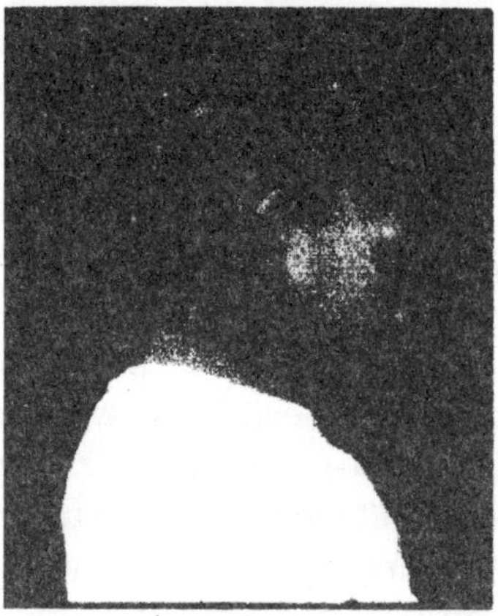

Fig. 8. A young boy suffering from serious ringworm infection, resulting in hair loss; regular application of neem oil cured the skin affliction in ~3 weeks. Neem oil also cures scabies, which is widespread among children in rural Africa *(Photo by R.C Saxena).*

By far the largest outlet for neem oil in India (one-fourth of~83,000 tons annually) is the soap industry which markets about a dozen different brands of bathing and laundry soaps. Neem oil also is used in the manufacture of dog soaps and shampoo that control, ticks, fleas, and lice. In Germany, it is used in herbal hair oil, hair tonic, and nail oil. A newly founded U.S. company is marketing a variety of toiletries based on neem oil and extracts.

Neem twigs are used daily by millions in Bangladesh, India, and Pakistan as disposable toothbrushes. Neem toothpaste and powder are reputed as effective dentrificc products. Extracts of neem bark are used in some toothpaste and mouthwashes singly or in combination with other herbal extracts. The inhibiting effect of aqueous neem extract upon bacteria influencing plaque formation has recently been confirmed. Other medicinal and veterinary uses of neem are emerging. Neem compounds and extracts may prove to be a rich cornucopia of many novel therapies.

Animal Care

Cattle leaf supplements containing neem leaf powder are worm killers. Creams containing neem oil are used for animal wound dressing and also act as fly and mosquito repellents. In trials conducted in Kenya, application of neem oil on cattle repelled all stages of *Rhipicephalus appendiculatus, Boophilus decoloratus,* and *Amblyomma variegatum* (Fig. 9). Also in Kenya, rural farming households have found in neem an inexpensive way to protect the native chicken from the dreaded New Castle virus disease. Neem seed powder or cake mixed in water and provided for drinking by chicken or chicken forced-fed on neem cake during epidemic survived, while others died.

Fig. 9. Healthy cattle grazing in a neem grove; note they are standing peacefully and not swishing their tails, indicating freedom from biting flies. Application of neem oil also repels cattle ticks and lice *(Photo by R.C. Saxena).*

Potential in Regulating Human Fertility

Today's exploding growth in human population is seriously depleting the world's natural reserves and economic resources. Unless the run-away human population growth rate is slowed down there would be little hope for raising everyone out of poverty in the developing world. Beside educational constraints, the non-availability of inexpensive methods of contraception, which do not cause trauma, nor violate aesthetic, cultural, and religious sensitivities of people, limit the success of birth regulation programmes. However, recent findings indicate that some neem derivatives may serve as affordable and widely available contraceptives. Oral feeding of extracts of neem leaves conferred antifertility in rats. Also, intravaginal administration of neem oil confers abortifacient and spermicidal activity. In rats, a single intra-uterme application of neem oil reversibly blocked fertility for 5 months, while a single intra-vas administration of neem oil rendered male rats infertile for 8 months. The treatments were non-inflammatory and did not affect libido. A neem-based polyherbal cream and pesseries with dual properties of contraception and alleviation of genital infections, caused by vaginal pathogens, has been developed in India. Recently, a pre-coital vaginal, neem-based contraceptive, "CONCEPT," has been developed in India and is already at large-scale clinical trial stages in Mysore and Hyderabad in India.

Neem for Mitigating Rural Poverty

Poverty is not necessarily the want of money or cash in hand. In a wider sense, it is the lack of options, whether it is the non-availability of fertilizer for crop cultivation or pesticides for crop protection, medical remedies for family welfare, safe contraceptives for regulating family size, fuel or firewood for cooking, timber for furniture or dwelling, or the availability of appropriate technology for restoring wastelands, or absence of income generation and employment opportunities. In all these respects, neem could be a 'panacea,' particularly in rural areas. In India, during the neem fruiting season in June-July, seed collection provides employment and income to unemployed women, children, and infirm people. With growing demand worldwide for neem seed, neem honey, and other neem products, there is substantial scope for establishing cottage industries, and other small-scale enterprises in rural areas of Asia and Africa where neem is widespread. Since agriculture is the staff of life in rural areas, enhanced agricultural productivity through the use of inexpensive neem products in pest management could significantly contribute to alleviation of poverty in rural areas.

Recent Developments on Neem in the Developed World

For nearly the past two decades, neem has come under close scientific scrutiny as a source of novel, natural insecticides in numerous international conferences, mostly held in developed countries; the latest being in Australia in Febbruary 1996. Nearly 2000 scientific papers have been published to date on neem. Australia with its large tracts of unused arid and semi-arid lands may become the largest grower of neem in the next couple of decades.

The interest in neem in the developed world is attributable to the fact that neem-based pest control products with diverse modes of action are not only effective against pests, but also inherently safer, less persistent in the environment, and less prone to the problem of pest resistance than the synthetics. Today, technical grade neem active ingredients, principally *azadirachtins, fetch the highest price, about US$ 375/kg* as compared with US$ 75/kg for pyrethrum (Isman 1995). In 1989, the use of 'Margosan,' containing 0.3 per cent azadirachtin, was granted approval from the US Environmental Agency for non-food uses on ornamentals and landscape plants; in 1993 EPA approved the use of neem products, such as "Neemix" on food crops. The U.S.-based W.R. Grace Co., which holds patents from the U.S. Patent and Trademarks Office on the method of extracting the insecticide from neem, is advertising Neemix as a *modern technology from ancient trees.* Agridyne, another U.S.-based company, is marketing 'Align' (with 3 per cent azadirachtin and 97 per cent inert ingredients, mainly other neem limonoids) for control of insect pests in vegetable, fruit, nut, and agronomic crops. Both products are now being used on commercial-scale crop protection in the USA. Neem seed extracts are being tested for forest insect pest management in Canada. Neem-based pesticides are expected to capture 10 per cent of the global pesticide market by the next decade. A technique using a neem extract as a fungicide also has been patented in the USA. Worldwide, nearly 50 patents have been granted on neem so far. The use of additives, adjuvants, activators, and even *Bt,* are being examined for potentiating the activity of azadirachtins against insect pests.

The patenting of neem pesticides and their formulations has evoked serious criticism and challenge in the developing world, particularly in India, as an example of 'folk wisdom piracy.' Efforts are underway in some European countries to extract azadirachtin on a commercial scale from neem calli. But, in all probability, such ventures would remain non-viable and economically unjustifiable. Quality neem seed with high aza content would remain the basic raw material for production of neem-based insecticides of the future. In that context, tropical countries of Asia and Africa could become major exporters of the raw material or even value-added finished products.

Conclusion

In the coming decades, the developing world would be facing four distressing crises, all counter-productive impacts of increased human activity and failure to use natural resources in a sustainable manner: 1. *Threat to food security* due to population pressure, 2. *Rural and urban poverty* and joblessness, 3. *Pollution* and degradation of arable land and water bodies, and 4. *Loss of biodiversity.* Neem has much to offer in solving global agricultural, public health, population, and environmental pollution problems. Certainly, it cannot be achieved without *building awareness* of its potential and *dissemination of neem-based technology* whether for pest management, public health, family welfare programmes, reforestation, or production and commercialization of various neem products for domestic use or exports. *More neem trees of superior ecotypes or genotypes will have to be grown,* particularly as a strategy for restoring degraded and marginal lands and making them productive and remunerative. Certainly, this would not happen overnight. If we are planning for long-term sustainability, then an investment in a time frame of 5 to 10 years is insignificant. Also, financial support, backed by favourable policies for neem promotion, production, and commercialization will be necessary.

As pointed out above, the demand for neem products, especially the seed as the basic raw material, is going to increase by leaps and bounds. Herein also lies a solution for creating income generation and job opportunities in rural and tribal areas. Neem-based industries in urban and industrialised regions would also create job slots for producing value-added products for domestic consumption and exports. Neem-based human fertility agents are almost on the anvil for stemming the tide of population explosion in developing countries. Compared with the existing methods of contraception, the acceptance of neem-based fertility regulating agents in Asia and Africa should be higher as neem is a household name and has a good track record of human safety and lack of side effects. Neem should also play a significant role in enriching the floral and faunal biodiversity as a huge variety of organisms, from insects to birds and mammals subsist on neem. Increased planting of neem along roadsides and avenues should make the cities more livable and make rural areas more attractive than now. Use of neem-based pest control agents and fertilizers should reduce pesticide-related hazards and pollution on land and in water bodies. In fact, the neem story is just unfolding. Tropical countries where neem can thrive have much to gain from an increased awareness of neem's hidden 'treasures'.

Selected References

Adhikary, S. 1985. Results of field trials to control the diamondback moth, *Plutella xylostella* L., in Togo by application of crude methanolic extracts of leaves and kernels of the neem tree, *Azadirachta indica* A. Juss, in Togo. *Z. Angew. Entomol.* 100:27-33.

Ascher, K.R.S. and J. Meisner. 1989. The effects of neem on insects affecting man and animal, pp. 113-131. In M. Jacobson (ed.), 1988 Focus on Phytochemical Pesticides. Vol. 1, The Neem Tree. CRC Press, Boca Raton.

Cobbinah, J.R. and K. Olei-Owusu. 1988. Effects of neem seed extracts on insect pests of eggplant, okra and cowpea. *Insect Sci. Appl.* 9:601-607.

Dreyer, M. 1987. Field and laboratory trials with simple neem products as protectants against pests of vegetables and field crops in Togo, pp. 431-447. In *Proc. 3rd. Int. Neem Cong., Nairobi, Jully 1986.* GTZ, Eschborn.

Dreyer, M. and C. Hellpap. 1991. Neem – a promising natural insecticide for small scale vegetable production in tropical and subtropical countries. *Z. PflKrankh. PflSchutz.* 98:428-437.

ICIPE (International Centre of Insect Physiology and Ecology). 1998. Natural pesticides from neem, pp. 93-110. *Annual Report,* 1995-1997. ICIPE, Nairobi.

Isman. M.B. 1995. Leads and prospects for the development of new botanical insecticides. *Rev. Pestic. Toxicol.* 3:1-20.

Kega, K. M. and R.C. Saxena. 1996. Neem derivatives for management of *Sitophilus zeamais* Motschiulsky [Coleoptera: Curculionidae]. *Paper presented at the Int. Neem Conf., Australia, 1996* (abstract).

Karel, A.K. 1989. Response of *Ootheca bennigseni* (Coleoptera: Chrysomelidae) to extracts from neem. *J. Econ. Entomol.* 82:1799-1803.

Musabyimana, T. and R.C. Saxena. 1999a. Efficacy of neem seed derivatives against nematodes affecting banana. *Phytoparasitica* 27:43-50.

Musabyimana, T. and R. C. Saxena. 1999b. Use of neem seed derivatives for sustainable banana pest management. Paper presented at the World Neem Conference, Vancouver, May 1999.

National Research Council. 1992. Neem – A Tree for Solving Global Problems. Report of an ad hoc panel of the Board on Science and Technology for International Development. National Academy Press, Washington DC, 141 p.

Rice, M. 1993. Development of neem research and industry in Australia, pp. 8-24. *In Souvenir, World Neem Conf., Bangalore, 1993.*

Ostermann, H. 1992. Zur Wirkung und Anwendung einfacher Niemprodukte in kleinbauerlichen Tomaten-, Vignabohnen- und Amaranthkulturen in Niger. Doctor, thesis., University of Giessen, Germany.

Parveen, G. and M.M. Alam. 1998. Efficacy of neem products for the management of root-knot nematodes on tomato, in soil polluted with heavy metals, cadmium and lead, pp. 235-244. In R.P. Singh and R.C. Saxena [Eds.], *Azadirachta indica* A. Juss. Oxford and IBH, New Delhi, India.

Rossner, J. and C.P.W. Zebitz. 1987. Effect of neem products on nematodes and growth of tomato *(Lycopersicon esculentum)* plants, pp. 611-621. In H. Schmutterer and K. R. S. Asched [Eds.], Natural Pesticides from the Neem Tree *(Azadirachta indica* A. Juss) and Other Tropical Plants, *Proc. 3rd. Int. Neem Conf.,* Nairobi, 1986. GTZ, Eschborn.

Saxena, R.C. 1989. Insecticides from neem, pp. 110-135. In J.T. Arnason, B.J.R. Philogene, and P. Morand (eds.), Insecticides of Plant Origin. *ACS Symp. Ser. 387*. American Chemical Society, Washington, DC.

Saxena, R.C. 1995. Pests of stored products, pp. 418-432. In H. Schmutterer (ed.), The Neem Tree, *Azadirachta indica* A. Juss., and Other Meliaceous Plants: Source of Unique Natural Products for Integrated Pest Management, Medicine, Industry and Other Purposes. VCH, Weinheim.

Saxena, R.C. 1998. "Green Revolutions" without blues: Botanicals for pest management, pp. 111-127. In G.S. Dhaliwal, N.S. Randhawa, R. Arora, and A.K. Dhawan [Eds.], Ecological Agriculture and Sustainable Development, Vol. 2. Indian Ecol. Soc, Ludhiana, India.

Saxena, R.C. and E.L. Kidiavai. 1997. Neem seed extract applications as low-cost inputs for management of the flower thrips in cowpea crop. Phytoparasitica 25:99-110.

Schmutterer, H. 1990. Properties and potential of natural pesticides from the neem tree, *Azadirachta indica. Annu. Rev. Entomol.* 35:271-297.

Schmutterer, H. 1995. The Neem Tree, *Azadirachta indica* A. Juss., and Other Meliaceous Plants: Sources of Unique Natural Products for Integrated Pest Management, Medicine, Industry and Other Purposes. VCH, Weinheim. 696 p.

Sharma, V.P., M.A. Ansari, and R.K. Razdan. 1993. Mosquito repellent action of neem *(Azadirachta indica)* oil. *J. Amer. Mosquito Control Assoc.* 9:359-360.

Siddig, S.A. 1987. A proposed pest management programme including neem treatments for combating potato pests in Sudan, pp. 449-459. In H. Schmutterer and K.R S. Ascher [Eds.], Natural Pesticides from the Neem Tree *(Azadirachta indica* A. Juss) and Other Tropical Plants. *Proc. 3rd Int. Neem Conf.*, Nairobi, 1986.

Siddig, S.A. 1988. Cultural means of controlling potato tuber moth *(Phthorimaea operculella* Zell.) and improvement of potato yield and quality in the Sudan. Acta Hort. 218:281-287.

Siddig, S.A. 1991. Evaluation of neem seed and leaf water extracts and powders for the control of insect pests in the Sudan. Tech. Bulletin # 6. Agric. Res. Corp., Shambat Res. Station. 39 p.

Talekar, N.S. and A.M. Shelton. 1993. Biology, ecology, and management of diamondback moth. *Annu. Rev. Entomol.* 38:275-301.

Vyas, B.N., N.B. Godrej, and K.B. Highway. 1996. Potential of Nimin-coated urea for enhancing nitrogen efficiency by crops. Paper presented at *Int. Neem Conf., Australia, 1996 (abstract).*

13

R&D and Quality Control of KVI Products

Dr. Naresh Pal

Preamble

Research and development is centred around the production of the product with better quality. Research is the backbone of any organisation for its proper growth and sustenance. All premier organisations of the country and abroad have a strong research base because it not only shows its strength but helps it sustaining in the changing global scenario of the advancement. More the research activities better the quality of the end product whether it be a manufacturing organisation or simply a service based organisation. This is becoming more and more important in the changing global market scenario as quality of the products as well as services is required with new parameters and standards. Thus, there is a strong need for the KVI Sector also to adopt market requirements for its products as well as services.

Sincere efforts are being made by KVIC to sustain in the global market by improving the quality of its products for which a strong research base by creating technical interfaces as well as quality network by giving accreditation to existing laboratories of national repute for testing and certification of KVI products throughout the country.

The Government of India has recently announced a KVI-Package where special emphasis is given for providing S&T input for improving the quality of the products of the KVI Sector. For strengthening of sector specific R&D autonomous societies are being created at different places. JBCRI, Wardha as a first step is being converted into Mahatma Gandhi Institute for Rural Industrialisation (MGIRI). Similarly the Khadi Gramodyog Prayog Samiti. (KGPS) will be developed further for research and development in Khadi. The Nasik Centre will be a society for research and development for soaps, detergents and essential oils. BHU, Varanasi will be developed for herbal and ayurvedic medicines. Five more centres are to be identified by the government.

For sector specific RD technical interfaces will be established in selected IITs, RECs, TTIs and major KVI institutions. For quality upgradation of KVI products services of national laboratories will be availed. All these efforts will lead to improved quality of KVI products.

Concept

Quality which is an important in product using modern method of Research and Development is actually an emotional experience for the customers who want to feel good about their purchases to feel that they have got the best value, know that their money has been well spent and take pride in their association with a company with a high quality image.

Though there is no universally accepted definition of quality the common elements involved with this are as under:

(a) Quality involves meeting customer expectations.

(b) Quality applies to products, services, people, processes and environments.

(c) Quality is an ever changing state.

The quality measures involve not only those contributing directly to the product quality but also activities that contribute indirectly to the same as sanitation, cleanliness, administration etc. The methods are written in documents called Standard Operation Procedures (SOPs).

Misconceptions about the quality:

- Very costly
- Massive paper generation

- Emphasis on correcting defects rather than preventing defects
- Super checking activity

The decentralised system of KVI Sector involves suppliers, manufacturers, certifying agencies, training agencies, academic institutes, professional organisations/industry associations, distributors and customer forums. Salient features of the decentralised system are:

- Agencies located throughout the country but scale of operation is very small.
- They are not aware about quality aspects, standardisation, documentation.
- They use alternative manufacturing processes to manufacture product.
- They can not afford to avail the services of quality testing labs.
- Their scale of operation is limited to the regional market only.

Quality system for the KVI Sector

For development of a specific plan for quality assurance in KVI sector following steps are needed:

- Documentation of all the procedures and processes
- Preparation of quality manual for each product.
- Identifying testing and certifying laboratories/agencies.
- Evolving procedure for availing the services of these agencies.
- Documentation of the procedures.
- Making QA mandatory for all KVI products.
- Timely review and monitoring.

Existing Status of Quality Assurance in KVI Products

At present FPO License is required for production of Food Products in KVI Sector but BIS and Agmark is not a must but some of the institutions are going for it. However for the recently launched Sarvodaya brand of KVIC both Agmark and BIS have been made compulsory.

Quality Certification Agencies

The village industry group covers around 117 activities for the purpose of implementation of VI programme. The articles produced under all these industrial activities have to go through various quality control measures under the acts established for the same which are as under:-

(i) Agmark Laboratory

(ii) F.P.O License

(iii) B.I. Standards

(iv) Drugs Act

All above acts cover different edible and non-edible commodities like Agmark Laboratory covers various spices, cereals, pickles etc. and B.I. Standard covers both edible and non-edible articles. KVIC is also a member of Bureau of Indian Standard for products under soap, oil, handmade paper, matches etc. All departmental Bhavans of KVIC are selling the products strictly under the norms of Agmark Laboratory and FPO License whereas the Bhavans established by the institutions do not follow this.

Most of the organisations producing food products both in organised and unorganised (KVI) sector though follow the standards prescribed by above acts but do not intend to go for license from them. The reason being once they are authorised for using their brands/stamps they may be penalised in case of any deviations from the set standards.

Objectives of Quality Control

(a) Wider choice for the customer.

(b) Assurance of quality.

(c) Increased earning for the artisans engaged in KVI sector.

(d) Making processed goods eco-friendly.

(e) Placement of 100 per cent eco-friendly cloth and other articles in domestic as well as export market.

Infrastructure available with KVIC

KVIC has got a very limited infrastructure available with it for quality assurance of KVI products which are detailed as under:

(i) Central Quality Control Laboratory, Nasik for quality assurance of soap and detergent products.

(ii) Central Bee Research and Training Institute, Pune for quality assurance of honey.

(iii) Gajanan Naik Palmgur Technical Development Centre, Dahanu for palm products.

(iv) Testing, Quality Control and Standardisation Unit, Dahanu for cottage match products.

(v) Dying and Printing Unit at C.B. Kora Institute of Village Industries, Borivali for chemical testing of Khadi Textiles.

(vi) Khadi Gramodyog Prayog Samiti, Ahmedabad for physical Testing of Khadi Textiles.

System of Quality Control for KVI Sector-Issues

Increase the Consciousness about Quality Assurance System

There is a need in KVI Sector to make the people involved in manufacturing and marketing process of KVI products to be more quality conscious. The KVI units should also be awared about the benefits of setting up of quality system in their units as the customers are much conscious about the quality aspects and need the approval of regulatory authority such as AGMARK, FPO, BIS, ISO 9000 etc. for the products.

Setting up of Packaging Design Centre

The attractive packaging is the need of the hours in the recent time for marketing of the products. There should be a design centre which can study and conduct survey among the customers, producers, end users etc. so that the attractive design of packaging can be developed.

Monitoring System

There should be a monitoring system for conforming the standard set for the products by the regulatory authority. The samples can be drawn starting from the raw material stage, processing stage, godown stage or market places. The responsibility of the conformity of the products to the set standards can be given to some independent agencies or can be taken up by technical staff of KVIC under the supervision of Industry Directors through State Directors.

Reporting System

There should be a established reporting system so that of the test report of the products can be send to all the concerned in KVI Sector besides the regulatory authority. The format of the reporting should be very simple and informative. There should be a system of analysis of the report and drawing the conclusion based on the reports.

Special Emphasis on the Quality of the thrust Area Product:

At present there are 3 types of thrust areas product:

Savodaya Products	Honey, pickel, soap, agarbatti, spices
Desi Aahar	Organic food items such as cereals, pulses, jaggary, honey, dalia, masala, etc.
Khadi Brand	The product of elite nature such as essential oil.

The manufacturing units, involved in any one of these three above mentioned products, should set up the minimum set of quality control system to ensure the quality of the product for the classified customers.

Infrastructure for Testing

For testing and maintaining quality of KVI products sufficient infrastructure is required which can be availed in two ways:

(a) Create new.

(b) Utilise infrastructure of CSIR, BIS, AGMARK.

To create own system/mechanism for quality control and quality assurance is costly so KVIC proposes to offer accreditation to various laboratories of:

(i) Govt. Sector: CSIR/I IT's/REC's/Universities/BIS/AGMARK/FPO

(ii) Pvt. Sector: SIIR, Delhi/Dabur/Ranbaxy etc.

Proposed Functions of Laboratories

(i) Testing/checking of KVI products, i.e.

1. Raw material
2. Intermediate
3. Finished products

(ii) Develop standards for products which are not covered so far by BIS/AGMARK/FPO etc.

1. These laboratories will create awareness about the quality concept among the KVI programme implementing agencies
2. These laboratories will cater to the needs of local KVI institutions with reference to their quality control requirements for
   - Raw material
   - Finished goods
   - Utilities
3. The laboratories will standardize the norms for quality of all KVI products, test the samples of production units and certify its quality.
4. These laboratories will have close interaction with the KVI units and also suggest suitable measures to improve and maintain the quality of the products.
5. The institutions/artisans working within the jurisdiction of these laboratories will ensure that they keep their products tested every time they start a particular batch of production

and sell it only after getting a certificate of quality from the concerned laboratory.

6. The charges for any kind of service rendered by these laboratories will have to be borne by the concerned KVI units.
7. These laboratories will also have an authority to have a surprise check of the production in process as well as that ready for dispatch. There will be a provision of penalty also in case at any stage discrepancy is observed in the quality of the end product in any manner.
8. The programme implementing agencies/artisans may approach KVIC for financial assistance for modification in the equipment/machinery etc. suggested by these laboratories for upgrading/maintaining quality of their products.

14

Potential of Tree Borne Oilseeds and Strategy to Harness Natural Wealth for Generation of Employment and Building National Wealth

K.D. Kamble

Among the perennial trees, while coconut and palm are perennial trees of agricultural origin, the tree borne oilseeds (TBOS) referred in this paper are perennial trees of forest origin. Some of them are in the forest areas while some like Neem, Karanja and Mohua are grown as avenue trees. Some annual plants which grow wild due to natural regeneration such as Thumb and by products such as Mango Stone and Cashew nut rejects are also included under this category.

The estimated potentiality of some important TBOS, based on the survey conducted by KVIC is shown in Table 1. The total seed potentiality of the 9 varieties of TBOS is around 6695 thousand tons. The total potentiality of 75 TBOS is yet to be arrived at. The TBOS vegetation being scattered and variable depending on the vagaries of nature, it has been very difficult to estimate the potentiality accurately.

While some of the TBOS are familiar to the commerce and industry still many of them are less known. To cite some examples Tumba Seed (Citrulus colocynthis (cucurbitaceae) possessing 21 per cent oil content, occurring in Rajasthan has the potential of around 23,000 tons. Ratnajyoth (Jahopa Curcas) possessing 45 per cent oil content in kernel, occurring in most of the states has the potential of around 30,000 tons. Chirandi (Litsea Umbrosa) possessing 55 per cent oil content in the kernel, occurring in Himalaya, Khasi Hills and Manipur has a potential of around 10,000 tons.

The seed (Fruit) falling season of these TBOS vary with the type of seed. Table 2 shows the fruiting season for 14 varieties of TBOS. The agencies engaged in seed collection activity should encourage the seed collectors to collect all varieties of TBOS available in a particular locality in order to provide the seed collector with a more sustained income rather than a seasonal income. While some TBOS such as neem are better stored as seed, some TBOS such as Karanja are better stored as kernel. Moisture plays a vital role in storage of the TBOS. For most variety of seeds the critical moisture content is around 8 to 10 per cent. Attack by fungus and weevilled in presence of high moisture is the main cause for deterioration in quality of these seeds while seed damage, foreign matters are other causes.

The fruit falling season of most of these TBOS coincide with the onset of monsoon, thus making it difficult to dry and store the seed in remote areas. Seeds such as Dhupa and rubber gets fully deteriorated within 2 days after falling on the ground during the rainy season if they are left on the ground. Khadi and Village Industries Commission (KVIC) has developed one tunnel air drier for drying such seeds. This is a long tunnel (60 feet) with a rail on which there is a battery of trolleys fitted with trays. An oil-fired burner with a blower is used to blow hot air inside the tunnel from one end. The seeds are filled in the trays and kept in the trolley and pushed inside the tunnel one after the other.

The people to whom these seeds are accessible are not very enthusiastic to collect them for their livelihood as the return they get for collecting and processing are much less than they get for equal labour in agriculture and industry. The people who have no other occupation and in particular, the tribals in forest areas only take up the seed collection work. The season generally coincides with the onset of monsoon when agricultural operations are in full swing. Naturally, collection of TBOS is not taken up to the full extent in agricultural areas.

At present 98 institutions/cooperative societies financed

by KVIC are engaged in seed collection activity. They make regular propaganda during the seed-collecting season covering the villages in their area of operation. They also announce their purchase rate for that season. Usually, temporary seed collection depots are set up by these institutions to purchase from the seed collectors. During this season I have even seen the tribals travelling from the hill tracks covering around 5 km. By walk with a head load of TBOS to reach the depots and sell their week's collection.

The seeds are transported from the depots to the central place where preprocessing is carried out. Here the seeds are graded and dried in drying platforms before storage. These centres are equipped with decorticators, siever and winnower for processing the seeds oil extraction is done by small portable oil extraction unit developed by KVIC or by small expeller units of 1 to 2 tons capacity. The units usually where the oil obtained is partly used for their own consumption and the remaining is sold.

Recently, KVIC has taken up the task of encouraging these units to aim at collection of premium quality oilseeds. In particular, simple directives are provided to the local people through these institutions to get intact neem seed of good quality free from aflotosum with high Azadirachitin content. These institutions also provide a premium rate for good quality neem seeds, the quality of which is judged by some simple field oriented tests. Thus the collector is encouraged to bring the dried neem seed instead of the wet fruit. The units are crushing the neem kernel instead of crushing the neem fruit/seed as done earlier. This fetches a better price for the neem oil and the neem cake.

The oil obtained is mostly used by these units for consumption in their own soap making units for both toilet soap and laundry soap and the surplus quantity is sold. Now KVIC is encouraging these institutions to collect good quality neem seed to tap export market. If this effort is fruitful the same process can be repeated on many other TBOS such as Karanja, Undi, Nahor which are drawing the attention due to their special non-glyceride components.

Role of ST in Development of Tree Borne Oilseeds

There is vast scope for RD inputs in various fields related to Neoils and Soap Industry and some of the areas of immediate interest which can improve the quality, performance and output of V.I. units and also will generate additional sustainable employment is given below:

1. Standardizing the procedure for collection and processing tree borne oilseeds. Presently most of the Neoilseeds are processed for use in soap making. However, recently oils such as neem, undi, karanja etc. are valued for their pest repellent/nematode/medicinal properties etc. Hence there is need to introduce suitable technology for separation of these active principles.
2. Development of cottage level acid slurry making unit.
3. Development of new formulation for products such as transparent soap, paper soap etc.
4. Introducing small capacity vacuum plodder for improved quality toilet soap.

Table 14.1
Potentiality of Tree Borne Oilseeds

Trees	*Seed potential (tonnes)*	*Oil yield (per cent) (average)*	*Oil potential (tonnes)*	*Nature of oil*	*Number of Trees (in lacks)*
Mahua	4,90,000	35	1,71,000	Hard	173.21
Neem	4,42,300	20	88,460	Semi-hard	157.10
Karanja	1,11,000	27	30,000	Semi-hard	69.37
Kusum	90,000	33	30,000	Hard	30.00
Sal	55,04,000	25	6,88,000	Hard	1234.30
Khakan	46,300	33	15,400	Nut oil	15.43
Kokum	2,000	40	800	Hard	1.67
Nahor	5,700	40	2,300	Semi-hard soft	1.71
Undi	3,800	60	2,300	Semi-hard soft	1.14
Total	**66,95,100**		**10,28,260**		**1683.93**

Table 14.2
Fruiting and Flowering Season of Tree Borne Oilseeds

Name of TBO seed	*Height of tree*	*Flowering season*	*Fruiting season*
Sal	18-30 mtr.	May-June	July-August
Mahua	Upto 70 ft.	Feb.-April	May-July
Neem	40-60 ft.	Feb.-May	May-August
Karanja	Upto 18 mtrs.	April-Sept.	Feb.-May
Rubber	60-100 ft.	Feb-March	Aug.-Nov.
Pilu	15-30 ft.	March-April	May-June
Kusum	30-35 ft.	Feb.-April	June-July
Dhupa	15 mtr.	Jan.-March	May-July
Nahor	12-15 ft.	Varies May-July October-March	Varies July-Sept. March-June
Undi	Upto 40 ft.	March-April	July-August
Kokum	15-18 mtrs.	Nov.-Feb.	April-May
Mango	15-40 ft.	Dec.-Jan.	April-May
Phulwara	15 mtrs.	April-May	June-July
Pisa	Upto 40 ft.	Nov.-Jan.	May-June

Science can be used for good purposes as well as for destruction. Nuclear energy is electricity and the production of safe and economical nuclear power is the main mandate of the Department of Atomic Energy. Nuclear energy has its application in medicine, agriculture and industry. But nuclear energy is also nuclear weapons and he mentioned about the devastation that can be caused by nuclear weapons. He is, of course, correct. And that is why we do not want anybody to threaten India with the use of nuclear weapons. The security environment in our neighbourhood had deteriorated so much by 1998 that Prime Minister Vajpayee decided that India should become a Nuclear Weapon State and the successful nuclear tests of May 1998 achieved this purpose. But he also said that we will not be the first to use nuclear weapons nor shall we ever use them against those who do not have nuclear weapons. All India wanted was credible minimum nuclear deterrence. The greatest advantage of recognized strength is that you don't have to use it; that is the principle of credible deterrence.

In the context of rural development, generation of employment, creation of jobs is the key issue – creation of jobs outside the primary agriculture and related sectors. Today two-thirds of Indians are in the farm sector in one way or another and unless there is a shift away from that for a substantial portion of them, there are limits to agricultural productivity that can be sustained. Today, in America, 2.5 per cent of the population takes care of their agricultural production and they have achieved high levels of agricultural productivity! So in addition to primary farm production, we have to promote food processing and other industrial activities at the rural level. Creation of this kind of jobs requires that the ST community must address itself more and more to the application of science to rural industrial– or, for that matter rural agricultural– development.

Agricultural Production Increase

This, I view in two dimensions. One is related to national production statistics. The other is related to local production increase to take people living there above the poverty line.

TIFAC (Technology Information, Forecasting and Assessment Council) is an institution of the Department of Science and Technology. Rural development is one of its core areas. TIFAC has launched various agriculture - related projects in Bihar, Eastern UP, Kancheepuram, Uttaranchal and other places. The main focus of these projects is the improvement of productivity of different crops in a sustainable manner, diversification of agriculture, entrepreneurship development for unemployed youth, etc. One important success story – where Prof S.K. Sinha, the former Director of IARI, has been providing the leadership is the increase in productivity of paddy and wheat in an area of Bihar by more than a factor of two within about two years of TIFAC intervention by simply optimizing the management practices. This increases the income, ability to take risks, and builds the resource base of farmers. TIFAC advises the farmers to diversify in crops so that they don't generate a glut for their products locally. Other examples are the revival of chick pea cultivation in Deoria district (in eastern UP) after 20 years, growing vegetables under greenhouse condition in Uttaranchal and growing of new vegetables like snake gourd and bitter gourd in Kancheepuram.

Other Activities of TIFAC

Beyond agriculture, there are many other activities of TIFAC related to rural development – in the Dairy sector; the development of a cost-effective diagnostic kit based on the ELISA technique by BAIF, Pune, for the diagnosis of diseases in cattle and poultry; briquetting and pelletisation of agro waste for fuel purposes; making blocks from coir pith–a waste material in the coconut belts of India for export; they use it in the horticulture industry abroad.

Other Institutions Involved in Rural Development

There are many scientific institutions working in areas related to rural development–some represented in this meeting. DBT has trained rural women in the Buithal village of Bassi, Jaipur, in the cultivation of spirulina and preparation of papad, laddu etc. with spirulina as one of the ingredients and has helped in development of floriculture in Guwahati, mushroom cultivation in West Bengal, Tamil Nadu and other places – these are all applications of biotechnology. During the Lucknow Science Congress, I visited the Central Institute of Medicinal and Aromatic Plants of CSIR and their work will help in rural development. The National Innovation Foundation of DST helps to develop the innovative ideas of people who have had no formal education. There is the Schumacher Centre in Delhi. The work of the Department of Atomic Energy in the disinfestations, hygienisation and shelf life extension of food products has great relevance in the context of rural development. Prime Minister Vajpayee has said that the Tenth Plan will place great emphasis on the farm sector and rural development. In the Lucknow Science Congress, he asked who but us will worry about the dryland crops. The time is ripe to increase the application of S&T to rural industries because so many scientists are interested. The question is how to add value to and improve the quality of rural products! How to downsize equipment to cater to the small cooperative sector, while maintaining quality and economic competitiveness. After all, one of the main purposes of Technology Development is to improve the Quality of Life, particularly of those people who are living in rural areas.

Conclusion

I have come here to learn. The application of science to rural industries–and to rural development in general, has to include the application of the highest levels of Technology–whether it is human or plant genomics or advanced materials–or the extension of established scientific and technological techniques through proper management practices.

Rural products have to be competitive with possible alternate products in price and quality. Here I look at quality in two different ways–Quality as such and quality improvement. When you are exporting or competing with an imported product, you have to reach international quality standards. If you build a nuclear reactor, a car or are manufacturing a drug-including a herbal product– you must match international quality standards. There are many other products–including, say, consumer products from rural industries–where our aim should be quality improvement in steps until we finally reach the level of the best international products.

We need synergy among institutions working in rural development. Another problem is this. Good technological and scientific ideas have been implemented, say, in one place and we want to spread them and take them to other places. Here I want to suggest two solutions. One is what I call 'Nucleation and Growth'. This is how crystals grow! If something is already implemented in one rural area, spreading by diffusion is easier. Nucleation requires leadership; leaders, in whom the people living around have faith. Spreading prosperity around such areas is relatively easier than starting the same activity in another place, unless you have an equally committed leader. Of course, this also should be done. For this, we need networks.

15

An Approach Towards Indian Hand Made Paper Industry for Rural Empowerment

Dr. D.N. Singh, Mr. Sanjay Singh and Mr. T. Chandrasekhar

Introduction

The word 'paper' actually comes from the word 'papyrus' and was invented in 4000 B.C. by Egyptians. Papyrus was woven mat of reeds and pounded together into a hard, thin sheet. Later on in history, the ancient Greeks used a kind of parchment made from animal skins for the same purpose. By 753 A.D. the Arabs started making paper and introduced a major innovation – using cotton as a papermaking fiber. The technology reached Spain, France, Germany and England. It was only in the middle of the 19th century when France discovered that paper can be made from wood.

Paper is one of the most essential commodities of modern life, particularly for the purposes of communication and other human uses. However, Indian paper consumption is low in international terms even though demand has been increasing steadily for four decades. In 1950, total demand was around, 120,000 tonnes but today it is about 3.6 million tonnes. Per capita consumption in India is approximately 3.5 kg in contrast to China's consumption of 14 kg, Malaysia's 61 kg, Thailand's 27 kg and Singapore's over 200 kg.

Handmade Paper

With rapid industrialization demand for paper and paper products have risen sharply. This has led to a serious ecological problem. To manufacture one tonne of paper board, paper mills require 2.5-3.0 tonnes of dry forest-based materials. In India, for several decades after independence, paper mills were liberally permitted to use forest stock from virgin forests, including bamboo, as raw material for the manufacture of paper. About 70 per cent of the total requirements of pulp for paper was met from forest-based raw materials, wood and bamboo; the remaining 30 per cent come from agricultural residues, including paddy straw, rags and waste paper.

As a result forest cover has already come down from 30 per cent in the fifties to around 16 per cent today. Because of this grim scenario, the very survival of several modern paper mills is at stake: a classic instance of unsustainable and ecologically indefensible development.

Now we are left with no other choice but to shift to alternative sources and that too in a fairly big way, if paper needs are to be met.

The depletion of forests is not the only problem associated with major paper mills. There are several others: the major chronic problem is the generation of effluents which are highly toxic due to the large quantities of chemicals used, in addition, the problem of air-polluting emissions as well.

It is in this context of large scale destruction of forest areas and the necessity of promoting an ecologically sound and sustainable technology along with providing low-cost inputs, that hand made paper industry should be seen.

The handmade-paper industry is a good example of a sustainable model of development which meets several desirable criteria: it is eco-friendly and provides rural employment for both men and women, usually near their homes, since the industry can be located near villages. The export market is also beckoning. It can be made from a variety of non-forestry-based resources. Such materials include cotton rags (in the form of tailor / hosiery cuttings), and small quantities of waste paper as well. Other agro fibres like jute, straw and banana tree can also be used for purposes of blending with the primary fibres for mottling effect etc.

Various kinds of handmade paper such as bond paper, drawing paper, superior card paper, filter paper/board, watermark paper etc., can be manufactured by units in small scale sector using easy techniques.

Status of Handmade Paper Industry in India

In India the history of handmade paper started in the year 1522 A.D. when emperor Babar came to India. Much of the industry was destroyed with the entry of foreign paper mills during the 18th and 19th centuries, under colonial rule.

It was under Mahatma Gandhi's inspiration that efforts were made in the thirties to revive the nearly-defunct industry. The revival was carried out under–the umbrella of the Khadi and Village Industries Association set up in 1934 under Gandhi's guidance and which was an important components of the Swadeshi movement. Many freedom fighters received training in the making of handmade-paper and they were instrumental in setting up handmade paper units in different parts of the country. Eventually, after India gained independence, the care of the industry and its research and development (R&D) was taken over by the present Khadi and Village Industries Commission. Today the handmade paper industry is a successful technology that can easily be replicated in any other part of the world.

Handmade paper making industry consumes large quantities of water (about 300 lt/tonne of paper) and is energy intensive. The industry should look out for better and improved machinery and processes where the consumption of water and power can be reduced and or reused and or recycled. The technology is fairly simple to operate and can be established easily in any part of the country, depending upon the economic considerations.

While in 1953, there were only 35 units all over the country with an annual production value worth Rs. 5 lakhs, today the number has reached to about 350 working units with an annual turnover about Rs. 100 crores, employing more than 15,000 people.

Raw Materials

The industry till this day depends mainly on cotton rags and waste paper. The cotton rag fibres are strong and ideal for simple technology prevalent in Handmade Paper Industry. However, its hydroscopic nature trends to absorb moisture from the atmosphere adversely affecting to some extent the dimensional stability, finish and the stiffness of the paper which are essentially required by the modern printing equipment like offset machines and computers.

Selected agro fibres can also be utilized either as blends in conventional rag fibres or as exclusive materials for specific varieties of paper in higher value segment. In our country there are million tons of agro waste products such as rice straw, etc. available annually. These wastes which are often burnt or left to rot in the field have been identified as a promising material for handmade paper. Tobacco stalks (classified as waste) can also be processed into high quality paper. The pulp yield from tobacco stalk is about 65 per cent and is higher than other agro wastes (roughly 20-30 per cent). Thus the scope of alternate raw material as a part substrate may be considered for different quality product applications.

Handmade Paper Making Process

The Process Involves following simple steps

Sorting and dusting: The raw material, waste cotton rags, is sorted by hand to remove foreign materials like plastics, iron and dust etc. that would contaminate the pulp.

Rag Chopping: The sorted cotton rags are put through a mechanized rag chopper that cuts the rags into small uniform sized pieces. In the past this process was done manually.

Beating: The chopped rags are converted into a fine pulp in a 'Hollander Beater'. Inert chemicals, like rosin soap and alum are added to give the paper the desired physical properties and the consistency required. When making coloured and or textured papers, the colour dyes and or textured materials like straw, hemp, grass, silk or jute fibers etc., are added during this process.

Sheet Formation: There are two method of sheet formation which are being used in India:

(i) **Lifting:** Lifting is the modern method. It employs the use of a steel water tank fitted with an apparatus to lift/lower the mould. The mould is clamped between two wooden deckles (frames) in a water tank. A measured quantity of the pulp (depending on the thickness of the paper required) is poured evenly into the mould. The mould is then shaken side to side horizontally (to ensure cross linking of the fibers – a unique characteristic of handmade paper) and raised mechanically to drain excess water. A sheet of pulp is formed over the mould which is taken out after unclamping the frame.

(ii) **Dipping:** Dipping is the traditional method. In this method the pulp is transferred from the beater into a masonry trough or vat. Depending on the thickness of the paper required, the pulp is diluted by mixing it with water. The mould, made of a wooden frame with a wire mesh or a bamboo stick mat is dipped by hand into the pulp. The frame is then shaken side to side and lifted out of the vat. A sheet of pulp is formed over the mould.

Nowadays a few handmade paper making units also use continuous rotating drum technology for making sheets.

Couching: The wet paper sheet is then transferred onto a cloth/felt piece of fabric, a couch which acts as a interleaf separating the wet sheets.

Pressing: A manual/hydraulic press the squeezes out the excess water from the sheets. This compresses the pulp adding strength to the fiber and facilitating the drying process.

Drying: Sheets are dried in the open air. For coloured

papers drying the shade. Though slower, is preferred in order to obtain uniform colour on both sides of the sheet.

Cleaning and Sizing: The sheets are then inspected for unwanted foreign matter. Small particles and dirt are removed manually with a sharp instrument. The cleaned sheets may then be coated with starch in order to make them blot-free.

Calendaring: Each sheet is laced between metallic plated and passed though spring loaded rollers in calendaring machine. This smoothens the paper and also enhances the gloss.

Cutting: At this point the calendared sheet still has a deckle edge. The sheet may remain uncut if required by the customer. However, typically, the sheets are machine cut to a standard and desired size and then packed.

The Various Machineries Presently Utilized Commonly are:

Hollander Beater: To make pulp by beating the cotton or waste paper inputs. It consists of a heavy, U-shaped trough in which a drum with multiple knives rotates rapidly to cut and chop the raw material.

Hydrapulper (Optional): Which makes pulp quickly, but can handle only paper soaked in water, not cloth.

Univat: A device needing only a little skill to lift sheets of paper with uniform thickness: a tank fitted with deckle and mould which transfers the formed sheet onto cotton felt by the process of couching.

Screw Press: An efficient version of an age-old mechanism to create high pressure and to squeeze out excess water from a stack of newly formed (very wet) paper.

Calendar Rollers: With two chilled cylinders rotating in opposite direction to give a smooth finish to the sheets of paper passed through them.

In Handmade Paper Industry on an average, the cost under various heads are as below:

(a) Raw material : 35 per cent

(b)	Chemicals	:	15 per cent
(c)	Wages	:	20 per cent
(d)	Power and fuel	:	10 per cent
(e)	Packing and forwarding	:	2.5 per cent
(f)	Repairs and replacements	:	2.5 per cent
(g)	Overheads	:	15 per cent

As we can see-raw materials constitute the largest percentage of cost in the making of hand made paper. So steps are needed to develop alternate sources of raw material.

Strengths, Capabilities and Vision for Indian Handmade Paper Industry

Today, the handmade paper industry manufactures a wide variety of paper, including drawing paper for artwork, permanent document paper, dark-coloured card sheets, deckled-edged stationery, greeting cards, carry bags, watermarked paper for certificates, filter paper, pads and insulation paper. Output is also used for file covers, duplicating paper and tissue paper. Products exported include carry bags, deckle - edged drawing paper and converted stationery. With technical improvements available in recent years, and despite its non-glazed appearance, *handmade paper can now even be used for four-colour reproduction on offset machines.* Once this takes place on a large scale, handmade paper will replace mill products like cartridge paper, and maplitho paper, etc.

With increasing industrialization and development the market demand is getting increasingly varied and sophisticated. There is growing preference for environment-friendly products worldwide and increasing demand for handmade paper products. Considering this the handmade paper industry has great future market potential. There are vast export opportunities, which should be explored.

Simultaneously the industry has to gear itself to respond to the growing sophistication of market demands. Accordingly, there is a need to constantly stress on the following:

- Awareness about the current market trends and demands.

- Diversification–regularly introduce new products into the market, as per demand.
- Standardization–constantly make improvements in the manufacturing process.
- Customization–develop new recipes tailored to manufacture customiz d requirement.
- Quality management–maintain strict control of the quality for every batch of paper produced.

Demand has already diversified into newer areas which hand made paper can cater to easily. Handmade papers have been made for printers, hand quality paper, customized scented paper etc. One of the recent trends in fabrics is introduction of paper clothing, i.e., cloth made of paper and bamboo. More and more of these materials are expected to find their way onto store shelves as time goes by.

A S.W.O.T. analysis of Indian handmade paper industry provided as follows:

Strengths

1. Abundance of indigenous raw materials.
2. Availability of creative crafts workforce.
3. Ease of entry into the business due to less capital investment and skills easy to learn.

Weaknesses

1. Market concentration in few countries.
2. Lack of technology for product conversion leading to low production capacity.

Opportunities

1. Untapped potential of growing preferences for environmentally friendly products worldwide.
2. Increasing demand for handmade paper and products.
3. Intensification of publicity and promotions.

4. Upgradation of technology for paper conversion and improving productivity.
5. Improving and continuous market based creative product design and development.
6. Enhancement of management skills.

Threats

1. Lack of appreciation of handmade paper and products in the domestic market (worth Rs. 3000 crores, an estimate made by All India Small Paper Mills Association, 2001).
2. Under-utilized, unviable units due to continuation of obsolete technology.
3. Unsatisfactory quality of product and inadequate trained/ skilled human resources.
4. Growing competition globally.

Vision

The future belongs to innovative products and the hand made paper industry has to gear itself to be technologically competitive to quickly respond to such market needs in order to capture major global share and sustain competitively.

Recommended Measures

A detailed study is necessary to pinpoint the exact upgradation necessary in different processes and machinery involved in making of handmade paper with main orientation towards:

- Shifting to use of alternate raw materials and common facility centre for pulping.
- Process engineering to ensure quality, productivity, safety and environmental aspects.
- Development of different energy efficient machines and alternate sources.
- Design, development and diversification/conversion of products.

- Utilization of locally available resources, including human resource.
- Mechanization to ensure safety, hygiene, quality and skill development activities and not repeat *not* to replace workforce by machine for higher income alone, and
- Creation of employment generation opportunities.

Conclusion

Now with commitment of government for a 'Developed Nation', thrust for rural development has become necessary. This calls for integrated rural development through ST interface in various sectors of Small Scale Industries. Surely, handmade paper industry will play a vital role in this direction. The regional development through this integrated way also leads to increased awareness in other segments like education, nutrition, hygiene, health and better livelihood of our rural cousins.

Can KVIC, select a cluster of villages in a loop to develop cottage industries at different peripheries of the loop to promote 'Rural Technology (RooT) Vikas'.

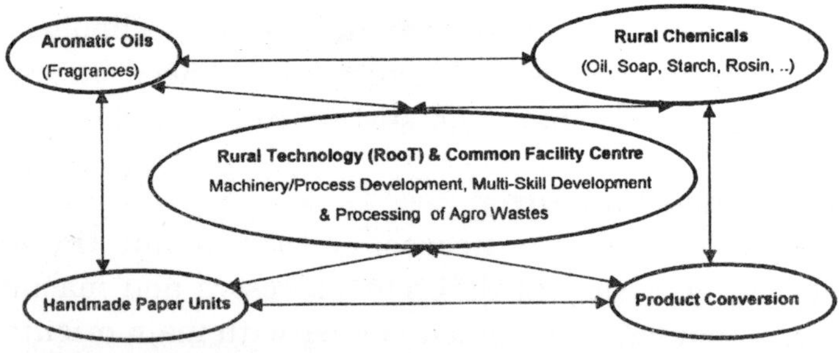

Acknowledgement: The authors are grateful to Dr. R. Chidambaram, Chairman-TIFAC for the guidance and encouragement given towards this programme. The authors are also grateful to Shri Laxmi Das, Chairman-Akhil Bharatiya Gramodyog Mahasangh for identifying the sector and giving us the opportunity to share our views. The authors are also thankful to Shri Pramod Garg, of Shankar Gramodyog Sewa Sansthan for the plant visit and providing valuable insights on this topic.

Appendices

Appendix A

Technologies Policy Statement, 1983

Preamble

Political freedom must lead to economic independence and the alleviation of the burden of poverty. We have regarded science and technology as the basis of economic progress. As a result of three decades of planning, and the Scientific Policy Resolution of 1958, we now have a strong agricultural and industrial base and a scientific manpower impressive in quality, numbers and range of skills. Given clear-cut objectives and the necessary support, our science has shown its capacity to solve problems.

The frontiers of knowledge are being extended at incredible speed, opening up wholly new areas and introducing new concepts. Technological advances are influencing life-styles as well as societal expectations.

The use and development of technology must relate to the people's aspirations. Our own immediate needs in India are the attainment of technological self-reliance, a swift and tangible improvement in the conditions of the weakest sections of the population and the speedy development of backward regions. India is known for its diversity. Technology must suit local needs and to make an impact on the lives of ordinary citizens, must give constant thought to even small improvements which could make better and more cost-effective use of existing materials and methods of work. Our development must be based on our own culture and personality. Our future depends on our ability to resist the imposition of technology which is obsolete or unrelated to our specific requirements and of policies which tie us to systems which serve the purposes of others rather than our own, and on our success in dealing with vested interests in our organizations: governmental, economic, social and even intellectual, which bind us to outmoded systems and institutions.

Technology must be viewed in the broadest sense, covering the agricultural and the services sectors along with the obvious manufacturing sector. The latter stretches over a wide spectrum ranging from village, small-scale and cottage industries (often based on traditional skills) to medium, heavy and sophisticated industries. Our philosophy of a mixed economy involves the operation of the private, public and joint sectors, including those with foreign equity participation.

Our directives must clearly define systems for the choice of technology, taking into account economic, social and cultural factors along with technical considerations; indigenous development and support to technology, and utilization of such technology; acquisition of technology through import and its subsequent absorption, adaptation and upgradation; ensuring competitiveness at international levels in all necessary areas; and establishing links between the various elements concerned with generation of technology, its transformation into economically utilizable form, the sector responsible for production (which is the user of such technology), financial institutions concerned with the resources needed for these activities, and the promotional and regulating arms of the Government.

This Technology Policy Statement is in response to the need for guidelines to cover this wide-ranging and complex set of inter-related areas. Keeping in mind the capital-scarce character of a developing economy it aims at ensuring that our available natural endowments, especially human resources, are optimally utilized for a continuing increase in the well-being of all sections of our people.

We seek technological advancement not for prestige or aggrandisement but to solve our multifarious problems and to be able to safeguard our independence and our unity. Our modernization, far from diminishing the enormous diversity of our regional traditions should help to enrich them and to make the ancient wisdom of our nation more meaningful to our people.

Our task is gigantic and calls for close co-ordination between the different departments of the Central and State Governments and also of those concerned, at all levels, with any sector of economic, scientific or technological activity, and, not least, the understanding and involvement of the entire Indian people. We look particularly to young people to bring a scientific attitude of mind to bear on all our problems.

Aims and Objectives

Aims

The basic objectives of the Technology Policy will be the development of indigenous technology and efficient absorption and adaptation of imported technology appropriate to national priorities and resources. Its aims are to:

(a) attain technological competence and self-reliance, to reduce vulnerability, particularly in strategic and critical areas, making the maximum use of indigenous resources;

(b) provide the maximum gainful and satisfying employment to all strata of society, with emphasis on the employment of women and weaker sections of society;

(c) use traditional skills and capabilities, making them commercially competitive;

(d) ensure the correct mix between mass production technologies and production by the masses;

(e) ensure maximum development with minimum capital outlay;

(f) identify obsolescence of technology in use and arrange for modernization of both equipment and technology;

(g) develop technologies which are internationally competitive, particularly those with export potential;

(h) improve production speedily through greater efficiency and fuller utilization of existing capabilities, and enhance the quality and reliability of performance and output;

(i) reduce demands on energy, particularly energy from non-renewable sources;

(j) ensure harmony with the environment, preserve the ecological balance and improve the quality of the habitat; and

(k) recycle waste material and make full utilization of by-products.

Self-Reliance

In a country of India's size and endowments, self-reliance is inescapable and must be at the very heart of technological development. We must aim at major technological break-throughs in the shortest possible time for the development of indigenous technology appropriate to national priorities and resources. For this, the role of different agencies will be identified, responsibilities assigned and the necessary linkages established.

Strengthening the Technology Base

Research and Development, together with science and technology education and training of a high order, will be accorded pride of place. The base of science and technology consists of trained and skilled manpower at various levels, covering a wide range of disciplines, and an appropriate institutional, legal and fiscal infrastructure. Consolidation of the existing scientific base and selective strengthening of thrust areas in it are essential. Special attention will be given to the promotion and strengthening of the technology base in newly emerging and frontier areas such as information and materials sciences, electronics and bio-technology. Education and training to upgrade skills are also of utmost importance. Basic research and the building of centres of excellence will be encouraged.

Skills and skilled workers will be accorded special recognition. The quality and efficiency of the technology generation and delivery systems will be continuously monitored and upgraded. All of this calls for substantial financial investments and also strengthening of the linkages between various sectors (educational institutions, R&D establishments, industry and governmental machinery).

Priorities

Need for Perspective Planning

The time scales involved in the generation of technology are long, even with imported elements. Therefore, relevant technologies in all areas of priority, particularly where large investments are to be made, should be clearly identified well in advance. The cost and time element involved in the import of technology and indigenous development will be given consideration. Components which could be assigned to the various institutions which are capable of developing them or which could be built up for such activities will be identified. Ministries concerned with large investments and production activities in areas such as food, health and energy will be provided with appropriate technical support through suitably structured S&T groups.

Employment

Human resources constitute our richest endowment. Conditions will be created for the fullest expression and utilization of scientific talent. Measures will be taken for the identification and diffusion of technologies that can progressively reduce the incidence of poverty and unemployment, and of regional inequalities. The application of science and technology for the improvement of standards of living of those engaged in traditional activities will be promoted, particularly household technologies. Technologies relevant to the cottage, village and small industries sector will be upgraded. In the decentralized sector labour must be diversified and all steps taken to reduce drudgery. In all sectors, the potential impact on employment will be an important criterion in the choice of technology.

Energy

Energy constitutes an expensive and sometimes scarce input. Therefore, the energy requirements both of a direct and indirect nature for each product and each production activity and the associated technology employed will be analysed. Measures will be devised to avoid wastage or non-optimal use of energy. Fiscal measures as necessary will be introduced to ensure these. Research and Development in the energy sector will aim at improving the efficiency of its production, distribution and utilization, as well as improvement of efficiency in processes and equipment.

Efficiency and Productivity

Technologies already employed will be evaluated on a continuing basis to realise maximum benefits in terms of increased production and lower costs, specially in the public sector enterprises. Every effort should be made to utilize by-products and wherever possible to recycle waste materials, especially those from urban areas. Programmes to make use of easily available and less costly materials will be supported.

Environment

Development should not upset the ecological balance for short as well as long-term considerations. Poorly planned efforts to achieve apparently rapid development, ignoring the long-term effect of many technologies on the environment, have resulted in serious ecological damage. It is, therefore, essential to analyse the environmental impact of the application of each technology. Due regard will be given to the preservation and enhancement of the environment in the choice of technologies. Measures to improve environmental hygiene will be evolved.

Some Specific Areas

In technology development special emphasis will be focused on food, health, housing, energy and industry. In particular, stress will be laid on: agriculture including dry-land farming; optimum use of water resources, increased production of pulses and oilseeds; provision of drinking water in rural areas, improvement of nutrition, rapid reduction in the incidence of blindness, eradication of the major communicable diseases (such as leprosy and tuberculosis), and population stabilization; low-cost housing; development and use of renewable non-conventional sources of energy; and industrial development

Indigenous Technology

Importance of Technology Development

Fullest support will be given to the development of indigenous technology to achieve technological self-reliance and reduce the dependence on foreign inputs, particularly in critical and vulnerable areas and in high value-added items in which the domestic base is strong. Strengthening and diversifying the domestic technology base are necessary to reduce imports and to expand exports for which international competitiveness must be ensured.

Inventions

The spirit of innovation and invention is the driving force behind all technological change. We must awaken our science and technology to the exciting challenges of our times, provide incentives to encourage inventors, and direct their efforts to areas of special importance. The system of rewards and incentives will be strengthened for inventions, innovations and technological breakthroughs and their utilization. The fullest opportunity will be provided to make use of inventions.

Enhancing Traditional Skills and Capabilities

Traditional skills and capabilities will need to be upgraded and enhanced, using knowledge and techniques generated by advances in science and technology. Technologies which will result in low-cost production and in products marketable close to the point of manufacture, particularly in the rural sector, will be promoted. Support will be given to technologies which reduce pressure on items in short supply and utilize improved local materials and methods. Government will give preference to products of such technologies in its own purchases. The adoption of technologies that can promote decentralized production will be helped through the support to design, marketing, quality control and other services.

Ensuring Timely Availability

The time cycle from scientific research to utilization is a long one. Hence the need to initiate action well in advance to identify and ensure timely availability and delivery of new technologies. Encouragement and support (fiscal, commercial and administrative) will be given to the production and user organizations to be associated with and participate in technology development efforts at appropriate stages.

Upgradation to Prevent Obsolescence

Technology is constantly on the move. The base of indigenous technology should be capable of utilizing world-wide advances and adapting them to local needs. The creation and strengthening of institutional structures for keeping track of international developments will receive urgent attention.

A strong central group will be constituted to undertake technology forecast and technology assessment studies and will inter alia draw up programmes of purposeful research. Arrangements will be made to provide high-level scientific advice in major sectors of the economy. Where big investments are involved or a large volume of production is envisaged, it will be incumbent on the Ministry or agency concerned to provide a technology forecast covering its requirements over a ten-year or longer period and evolve a strategy for development based on priorities.

Increasing the Demand for Indigenous Technology

Our country has already invested significant amounts in setting up research and development facilities as well as design consultancy and engineering capabilities. The technological potential inherent in this system of interlinked capabilities must be fully utilized, and in turn provide a fillip for further development from within the system. Incentives will, therefore, be provided to users of indigenously developed technology, and for products and processes resulting for such use.

Preferential Treatment

In view of the cost of technology development and the time necessary for successful marketing of a new or improved product, indigenously developed items are invariably at a disadvantage compared with imported products or those based on imported technologies and brand names. Support must therefore be provided through fiscal and other measures, for a limited period, in favour of products made through indigenously developed technologies, care being taken to ensure quality.

Fiscal Incentives

Suitable financial mechanisms will be established to facilitate investment on pilot plants, process demonstration units and prototype development in order to enable rapid commercial exploitation of technologies developed in laboratories. Linkages between scientific and technological institutions and development banks will be strengthened. Gaps in technology will be identified and suitable corrective measures taken with adequate allocation of resources. Fiscal incentives will be provided in particular to: promote inventions; increase the use of indigenously developed technology; enhance in-house Research and Development in industry; and efforts directed to absorb and adapt imported technology.

Design Engineering

Capabilities in design engineering are essential for the translation of know-how to commercial production. This is particularly important in areas relating to: agricultural production; agro-industries; metallurgical, chemical and petrochemical processes; machine tools; industrial machinery and capital goods; as well as for the construction and erection of entire plants. Building up and enhancing these capabilities will have a catalytic beneficial impact on the utilization of indigenous efforts that have resulted in product and process know-how. Existing design engineering capabilities will be strengthened and upgraded, and interaction encouraged between design engineering organizations, academic and research institutions and industry. Wherever gaps exist, design engineering capabilities will be developed and nurtured.

Engineering Consultancy

Engineering consultancy is a vital area for ensuring speedy technological and industrial development. It ensures the appropriate utilization of indigenous materials, plant and machinery. Engineering consultancy provides an essential link between R&D institutions and industry, and thus promotes effective transfer of technology. Capability for total systems engineering, process development and project management should be developed with collaboration if required. Wherever capability exists, utilization of Indian consultancy engineering organizations will be promoted. Even where foreign technical collaboration or consultancy is considered unavoidable, association of designated

Indian consulting engineering organizations would be preferred. Indigenous engineering consultancy, in both private and public sectors, will be promoted on a sound professional basis in the context of the overall national perspective of technological self-reliance.

In-house R&D

In-house R&D units in industry provide a desirable and essential interface between efforts within the national laboratories and the educational sector as well as production in industry. Appropriate incentives will be given to the setting up of R&D units in industry and for industry including those on a cooperative basis. Enterprises will be encouraged to set up R&D units of a size to permit the accomplishment of major technological tasks.

Technology Acquisition

Mix of Indigenous and Imported Technology

A policy directed towards technological self-reliance does not imply technological self-sufficiency. The criterion must be national interest. Government policy will be directed towards reducing technological dependence in key areas.

Advantage should be taken of technological developments elsewhere. This can also be achieved through well-defined collaborative arrangements in research and development.

At any given point of time, there will be a mix of indigenous and imported technology. However, technology acquisition from outside shall not be at the expense of national interest. Indigenous initiative must receive due recognition and support.

In the acquisition of technology, consideration will be given to the choice and sources of technology, alternative means of acquiring it, its role in meeting a major felt need, selection and relevance of the products, costs, and related conditions. A National Register on Foreign Collaboration will be developed to provide analytical inputs at various stages of technological acquisition.

Principles of Acquisition and Technology Assessment

Where the need to import technology is established, every effort should be made to ensure that it is of the highest level, consistent with requirements and resources. The technology import will be so planned as to have effective transfer of basic knowledge (know-why) and to facilitate further advancement.

Where the import of technology is contemplated, the level to which technology has been developed, or is in current use, within the country, shall be first evaluated. Lists of technologies that have been adequately developed to the extent that import is unnecessary will be prepared and periodically updated; in such areas no import of technology would normally be permitted; and the onus will be on the seeker of foreign technology, be it industry or a user Ministry, to demonstrate to the satisfaction of the approval authority that import is necessary.

Technology assessment systems will be reviewed. A technology assessment mechanism consisting of competent groups will render advice in all cases of technology import relating to highly sophisticated technology, large investments and national security. Aspects of employment, energy, efficiency and environment will be kept in view.

The basic principles governing the acquisition of technology will be:-

Import of technology, and foreign investment in this regard, will continue to be permitted only on a selective basis where : need has been established; technology does not exist within the country; the time taken to generate the technology indigenously would delay the achievement of development targets.

Government may, from time to time, identify and notify such areas of high national priority, in respect of which procedures would be simplified further to ensure timely acquisition of the required technology.

There shall be a firm commitment for absorption, adaptation and subsequent development of imported know-how through adequate investment in Research and Development to which importers of technology will be expected to contribute.

Unpackaging

Technology to fulfil a particular need consists of many components. It is necessary to develop capability to break down the total package of technology required for a purpose into components, some of which may be readily available or could be indigenously developed, and other that will need to be imported. Norms and guidelines for such unpackaging will be evolved.

Absorption of Technology

There shall be a commitment to ensure an adequate scale of investment in R&D for the absorption, adaptation and, wherever possible, improvement on and generation of new technology, making fullest use of overall national capabilities. Only thus can self-reliance be ensured and a technology generation process established firmly. Appropriate mechanisms will be evolved at the stage of technology assessment to ensure the absorption of imported technology.

Technological Information

The availability of an efficient system of collection and analysis of relevant technological information, including cost and other economic aspects, is a prerequisite for the appropriate choice of technologies. This will considerably enhance the possibility of obtaining favourable terms and conditions in acquisition of technology. Such a technology information base will be established.

Technology Transfer

Diffusion

Special efforts need to be made for the diffusion of technology in use to all beneficiaries who can employ them optimally. Appropriate measures shall be evolved to facilitate technology diffusion, including : horizontal transfer; technological support for ancillaries from large units; technology inputs to small units; and upgradation of traditional skills and capabilities.

International Competitiveness and Technology Exports

It is necessary to maintain international competitiveness in products, services and technologies that have export potential. Conditions for the marketing of indigenous technology and of products based on it will be improved. It is important in all such cases to conform to the highest international standards.

Technical Cooperation among Developing Countries

A concerted effort will be made to participate fully in technical cooperation among developing countries. Encouragement will be provided for participation in technology development programmes with other developing countries which can contribute to mutual national development.

Protection : Legislative Framework

Development of technology calls for large investments and often involves considerable risk. Encouragement will be given to obtaining necessary protection in all cases of indigenous technology development. A mechanism will be set up to ensure that national interests arising from the generating of technology are fully protected internationally in terms of industrial property rights.

Implementation

The success of the Technology Policy and the speed with which the various facets of the policy are implemented will depend to a considerable extent on a system for efficient monitoring, review and guidance and a scheme of incentives and disincentives.

Government will evolve instruments for the implementation of this Technology Policy and spell out in detail guidelines for Ministries and agencies of Government as well as for industries and entrepreneurs.

Success in implementation demands a conscious integrated approach covering technology assessment, development, acquisition, absorption, utilization and diffusion, and connected aspects of financing, based on overall national interests, priorities and the attainment of the most challenging technological goals.

Above all, the entire population must be imbued with self-confidence and pride in national capacity.

Indian Science and Technology must unlock the creative potential of our people and help in building the India of our dreams.

Appendix B

Science Policy: 1958

1. No. 131/CF/57. The key to national prosperity, apart from the spirit of the people, lies, in the modern age, in the effective combination of three factors, technology, raw materials and capital, of which the first is perhaps the most important, since the creation and adoption of new scientific techniques can, in fact, make up for a deficiency in natural resources, and reduce the demands on capital. But technology can only grow out of the study of science and its applications.
2. The dominating feature of the contemporary world is the intense cultivation of science on a large scale, and its application to meet a country's requirements. It is this, which, for the first time in man's history, has given to the common man in countries advanced in science, a standard of living and social and cultural amenities, which were once confined to a very small privileged minority of the population. Science has led to the growth and diffusion of culture to an extent never possible before. It has not only radically altered man's material environment, but, what is of still deeper significance, it has provided new tools of thought and has extended man's mental horizon. It has thus influenced even the basic values of life, and given to civilization a new vitality and a new dynamism.
3. It is only through the scientific approach and method and the use of scientific knowledge that reasonable material and cultural amenities and services can be provided for every member of the community, and it is out of a recognition of this possibility that the idea of a welfare state has grown. It is characteristic of the present world that the progress

towards the practical realisation of a welfare state differs widely from country to country in direct relation to the extent of industrialisation and the effort and resources applied in the pursuit of science.

4. The wealth and prosperity of a nation depend on the effective utilisation of its human and material resources through industrialisation. The use of human material for industrialisation demands its education in science and training in technical skills. Industry opens up possibilities of greater fulfilment for the individual. India's enormous resources of manpower can only become an asset in the modern world when trained and educated.

5. Science and technology can make up for deficiencies in raw materials by providing substitutes, or, indeed, by providing skills which can be exported in return for raw materials. In industrialising a country, heavy price has to be paid in importing science and technology in the form of plant and machinery, highly paid personnel and technical consultants. An early and large scale development of science and technology in the country could therefore greatly reduce the drain on capital during the early and critical stages of industrialisation.

6. Science has developed at an ever-increasing pace since the beginning of the century, so that the gap between the advanced and backward countries has widened more and more. It is only by adopting the most vigorous measures and by putting forward our utmost effort into the development of science that we can bridge the gap. It is an inherent obligation of a great country like India, with its traditions of scholarship and original thinking and its great cultural heritage, to participate fully in the march of science, which is probably mankind's greatest enterprise today.

7. The Government of India have accordingly decided that the aims of their scientific policy will be:

(1) to foster, promote, and sustain, by all appropriate means, the cultivation of science, and scientific research in all its aspects — pure, applied, and educational;

(2) to ensure an adequate supply, within the country, of research scientists of the highest quality, and to recognize their work as an important component of the strength of the nation;

(3) to encourage, and initiate, with all possible speed, programmes for the training of scientific and technical personnel, on a scale adequate to fulfil the country's needs in science and education, agriculture and industry, and defence;

(4) to ensure that the creative talent of men and women is encouraged and finds full scope in scientific activity;

(5) to encourage individual initiative for the acquisition and dissemination of knowledge, and for the discovery of new knowledge, in an atmosphere of academic freedom;

(6) and, in general, to secure for the people of the country all the benefits that can accrue from the acquisition and application of scientific knowledge.

The Government of India have decided to pursue and accomplish these aims by offering good conditions of service to scientists and according them an honoured position, by associating scientists with the formulation of policies, and by taking such other measures as may be deemed necessary from time to time.

Appendix C

Science and Technology Policy: 2003

A-Preamble

Science and technology have profoundly influenced the course of human civilization. Science has provided us remarkable insights into the world we live in. The scientific revolutions of the 20th century have led to many technologies, which promise to herald wholly new eras in many fields. As we stand today at the beginning of a new century, we have to ensure fullest use of these developments for the well being of our people.

Science and technology have been an integral part of Indian civilization and culture over the past several millennia. Few are aware that India was the fountainhead of important foundational scientific developments and approaches. These cover many great scientific discoveries and technological achievements in mathematics, astronomy, architecture, chemistry, metallurgy, medicine, natural philosophy and other areas. A great deal of this travelled outwards from India. Equally, India also assimilated scientific ideas and techniques from elsewhere, with open-mindedness and a rational attitude characteristic of a scientific ethos. India's traditions have been founded on the principles of universal harmony, respect for all creation and an integrated holistic approach. This background is likely to provide valuable insights for future scientific advances. During the century prior to Independence, there was an awakening of modern science in India through the efforts of a number of outstanding scientists. They were responsible for great scientific advances of the highest international caliber.

In the half century since Independence, India has been committed to the task of promoting the spread of science. The key role of technology as an important element of national development is also well recognised. The Scientific Policy Resolution of 1958 and the Technology Policy Statement of 1983 enunciated the principles on which the growth of science and technology in India has been based over the past several decades. These policies have emphasized self-reliance, as also sustainable and equitable development. They embody a vision and strategy that are applicable today, and would continue to inspire us in our endeavours.

With the encouragement and support that has been provided, there is today a sound infrastructural base for science and technology. These include research laboratories, higher educational institutions and highly skilled human resource. Indian capabilities in science and technology cover an impressive range of diverse disciplines, areas of competence and of applications. India's strength in basic research is recognized internationally. Successes in agriculture, health care, chemicals and pharmaceuticals, nuclear energy, astronomy and astrophysics, space technology and applications, defense research, biotechnology, electronics, information technology and oceanography are widely acknowledged. Major national achievements include very significant increase in food production, eradication or control of several diseases and increased life expectancy of our citizens.

While these developments have been highly satisfying, one is also aware of the dramatic changes that have taken place, and continue to do so, in the practice of science, in technology development, and their relationships with, and impact on, society.

Particularly striking is the rapidity with which science and technology is moving ahead. Science is becoming increasingly inter- and multi-disciplinary, and calls for multi-institutional and, in several cases, multi-country participation. Major experimental facilities, even in several areas of basic research, require very large material, human and intellectual resources. Science and technology have become so closely intertwined, and so reinforce each other that, to be effective, any policy needs to view them together. The continuing revolutions in the field of information and communication technology have had profound impact on the manner and speed with which scientific information becomes available, and scientific interactions take place.

Science and technology have had unprecedented impact on economic growth and social development. Knowledge has become a source of economic might and power. This has led to increased restrictions on sharing of knowledge, to new norms of intellectual property rights, and to global trade and technology control regimes. Scientific and technological developments today also have deep ethical, legal and social implications. There are deep concerns in society about these. The ongoing globalisation and the intensely competitive environment have a significant impact on the production and services sectors.

Because of all this, our science and technology system has to be infused with new vitality if it is to play a decisive and beneficial role in advancing the well being of all sections of our society. The nation continues to be firm in its resolve to support science and technology in all its facets. It recognizes its central role in raising the quality of life of the people of the country, particularly of the disadvantaged sections of society, in creating wealth for all, in making India globally competitive, in utilizing natural resources in a sustainable manner, in protecting the environment and ensuring national security.

B-Policy Objectives

Recognizing the changing context of the scientific enterprise, and to meet present national needs in the new era of globalisation, government enunciates the following objectives of its Science and Technology Policy:

- To ensure that the message of science reaches every citizen of India, man and woman, young and old, so that we advance scientific temper, emerge as a progressive and enlightened society, and make it possible for all our people to participate fully in the development of science and technology and its application for human welfare. Indeed, science and technology will be fully integrated with all spheres of national activity.
- To ensure food, agricultural, nutritional, environmental,

water, health and energy security of the people on a sustainable basis.

- To mount a direct and sustained effort on the alleviation of poverty, enhancing livelihood security, removal of hunger and malnutrition, reduction of drudgery and regional imbalances, both rural and urban, and generation of employment, by using scientific and technological capabilities along with our traditional knowledge pool. This will call for the generation and screening of all relevant technologies, their widespread dissemination through networking and support for the vast unorganized sector of our economy.
- To vigorously foster scientific research in universities and other academic, scientific and engineering institutions; and attract the brightest young persons to careers in science and technology, by conveying a sense of excitement concerning the advancing frontiers, and by creating suitable employment opportunities for them. Also to build and maintain centres of excellence, which will raise the level of work in selected areas to the highest international standards.
- To promote the empowerment of women in all science and technology activities and ensure their full and equal participation.
- To provide necessary autonomy and freedom of functioning for all academic and R&D institutions so that an ambience for truly creative work is encouraged, while ensuring at the same time that the science and technology enterprise in the country is fully committed to its social responsibilities and commitments.
- To use the full potential of modern science and technology to protect, preserve, evaluate, update, add value to, and utilize the extensive knowledge acquired over the long civilizational experience of India.
- To accomplish national strategic and security-related objectives, by using the latest advances in science and technology.

- To encourage research and innovation in areas of relevance for the economy and society, particularly by promoting close and productive interaction between private and public institutions in science and technology. Sectors such as agriculture (particularly soil and water management, human and animal nutrition, fisheries), water, health, education, industry, energy including renewable energy, communication and transportation would be accorded highest priority. Key leverage technologies such as information technology, biotechnology and materials science and technology would be given special importance.
- To substantially strengthen enabling mechanisms that relate to technology development, evaluation, absorption and upgradation from concept to utilization.
- To establish an Intellectual Property Rights (IPR) regime which maximises the incentives for the generation and protection of intellectual property by all types of inventors. The regime would also provide a strong, supportive and comprehensive policy environment for speedy and effective domestic commercialisation of such inventions so as to be maximal in the public interest.
- To ensure, in an era in which information is key to the development of science and technology, that all efforts are made to have high-speed access to information, both in quality and quantity, at affordable costs; and also create digitized, valid and usable content of Indian origin.
- To encourage research and application for forecasting, prevention and mitigation of natural hazards, particularly, floods, cyclones, earthquakes, drought and landslides.
- To promote international science and technology cooperation towards achieving the goals of national development and security, and make it a key element of our international relations.
- To integrate scientific knowledge with insights from other disciplines, and ensure fullest involvement of scientists and technologists in national governance so that the spirit and

methods of scientific enquiry permeate deeply into all areas of public policy making.

It is recognized that these objectives will be best realized by a dynamic and flexible Science and Technology Policy, which can readily adapt to the rapidly changing world order. This Policy, reiterates India's commitment to participate as an equal and vigorous global player in generating and harnessing advances in science and technology for the benefit of all humankind.

C. Strategy and Implementation Plan

Keeping in view these broad objectives, it is essential to spell out an implementation strategy that will enable identification of specific plans, programmes and projects, with clearly defined tasks, estimates of necessary resources, and time targets. Some of the key elements of the implementation strategy will be as follows:

Science and Technology Governance and Investments

Suitable mechanism will be evolved by which independent inputs on science and technology policy and planning are obtained on a continuous basis from a wide cross-section of scientists and technologists. It will utilize the academies and specialized professional bodies for this purpose. These inputs will form an integral part of the planning and implementation of all programmes relating to science and technology, as also in government decision making and formulation of policies in socio-economic sectors.

A greater integration of the programmes in socio-economic sectors with R&D activities will go a long way in ensuring a wider, more visible and tangible impact. This will call for a certain percentage of the overall allocation of each of the socio-economic ministries to be devoted for relevant programmes and activities in science and technology. The States will also be encouraged and assisted in the use of science and technology for developmental purposes through mechanisms set up for this, and in establishing linkages with national institutions for solving their regional and locale-specific problems.

A concerted strategy is necessary to infuse a new sense of dynamism in our science and technology institutions. The science departments, agencies and other academic institutions, including universities, i.e., the science and technology system as a whole, would be substantially strengthened, given full autonomy and flexibility, and debureaucratized.

Mechanisms will be established to review on a continuous basis the academic and administrative structures and procedures in the science and technology system at all levels, so that reforms could be effected to meet the challenges of the changing needs.

It will be ensured that all highly science-based Ministries/ Departments of Government are run by scientists and technologists. All the major socio-economic Ministries will have high-level scientific advisory mechanisms.

Government will ensure continued existence of an Apex S&T Advisory Body which will assist in formulating and implementing various programmes and policies. It will have appropriate representation of industry leaders, leading scientists and technologists and various scientific departments.

Government will make necessary budgetary commitments for higher education and science and technology. It will, through its own resources and also through contribution by industry, raise the level of investment to at least 2 per cent of GDP on science and technology by the end of the Tenth Plan. For this, it is essential for industry to steeply increase its investments in RD. This will enable it to be competitive, achieve greater self-reliance and self-confidence, and fulfill national goals.

Optimal Utilization of Existing Infrastructure and Competence

Science and technology is advancing at a very fast pace, and obsolescence of physical infrastructure, as also of skills and competence, take place rapidly. Steps will be taken to network the existing infrastructure, investments and intellectual strengths, wherever they exist, to achieve effective and optimal utilization, and constantly upgrade them to meet changing needs.

Strengthening of the Infrastructure for Science and Technology in Academic Institutions

A major initiative to modernize the infrastructure for science and engineering in academic institutions will be undertaken. It will be ensured that all middle and high schools, vocational and other colleges will have appropriately sized science laboratories. Science, engineering and medical departments in academic institutions and universities and colleges will be selected for special support to raise the standard of teaching and research. To begin with, a significant number of academic institutions, specially the universities, as also engineering and medical institutions, would be selected for this support to make an impact. Flexible mechanisms for induction of new faculty in key areas of science would be developed. Constancy of support and attention will be ensured over at least a ten-year period.

New Funding Mechanisms for Basic Research

The setting up of more efficient funding mechanisms will be examined, either by creating new structures or by strengthening or restructuring the existing ones, for promotion of basic research in science, medical and engineering institutions. In particular, administrative and financial procedures will be simplified to permit efficient operation of research programmes in diverse institutions across the country.

Creation of world class facilities in carefully selected and nationally relevant fields will be undertaken, to enhance our international competitiveness in areas where we have strengths, opportunities or natural advantages. Indigenous expertise will be used to the maximum extent possible. This would help in nurturing high quality talent and expertise in experimental science and engineering.

Human Resource Development

The number of scientists and technologists, while being large in absolute numbers, is not commensurate with the requirements in quality and when measured on a per capita basis. The demand is bound to increase in the coming years with more intensive activities involving science and technology. There is need to progressively increase the rate of generation of high quality skilled human resource at all levels. This process would naturally entail reversing the present flow of talent away from science, by initiating new and innovative schemes to attract and nurture young talent with an aptitude for research, and by providing assured career opportunities in academia, industry, government or other sectors. *In order to encourage quality and productivity in science and technology, mobility of scientists and technologists between industry, academic institutions and research laboratories will be ensured.*

For building up the human resource base in relevant areas, the agencies and departments concerned with science and technology will make available substantial funding from their allocation. Flexible mechanisms will be put in place in academic and research institutions to enable researchers to change fields and bring new inputs into traditional disciplines, and also to develop inter-disciplinary areas. There will be emphasis on a continuing process of retraining and reskilling to keep pace with the rapid advances taking place. Wherever considered necessary, training abroad will be resorted to, so as to build up a skilled base rapidly.

Women constitute almost half the population of the country. They must be provided significantly greater opportunities for higher education and skills that are needed to take up R&D as a career. For this, new procedures, and flexibility in rules and regulations, will be introduced to meet their special needs.

New mechanisms would be instituted to facilitate the return of scientists and technologists of Indian origin to India, as also their networking, to contribute to Indian science and technology.

Schemes for continuing education and training of university and college teachers in contemporary research techniques and in emerging areas of science will be strengthened and new innovative programmes started.

It will also be ensured that higher education is available to the widest possible section of creative students, transcending social and economic barriers.

Technology Development, Transfer and Diffusion

A strong base of science and engineering research provides a crucial foundation for a vibrant programme of technology development. Priority will be placed on the development of technologies which address the basic needs of the population; make Indian industries—small, medium or large—globally competitive; make the country economically strong; and address the security concerns of the nation. Special emphasis will be placed on equity in development, so that the benefits of technological growth reach the majority of the population, particularly the disadvantaged sections, leading to an improved quality of life for every citizen of the country. These aspects require technology foresight, which involves not only forecasting and assessment of technologies but also their social, economic and environmental consequences.

The growth rate in productivity of the Indian economy has been below its true potential, and the contribution to it of technological factors is inadequate. Similarly, Indian exports today derive their comparative advantage through resource and labour rather than through the power of technological innovation. The transformation of new ideas into commercial successes is of vital importance to the nation's ability to achieve high economic growth and global competitiveness. Accordingly, special emphasis will be given not only to R&D and the technological factors of innovation, but also to the other equally important social, institutional and market factors needed for adoption, diffusion and transfer of innovation to the productive sectors.

Intensive efforts will be launched to develop innovative technologies of a breakthrough nature; and to increase our share of high-tech products. Aggressive international benchmarking will be carried out. Simultaneously, efforts will be made to strengthen traditional industry so as to meet the new requirements of competition through the use of appropriate science and technology. This industry is particularly important as it provides employment at lower per capita investment, involves low energy inputs, and carries with it unique civilizational traditions and culture. Value addition, and creation of wealth through reassessment, redistribution and repositioning of our intellectual, capital and material resource will be achieved through effective use of science and technology.

Deriving value from technology-led exports and export of technologies will be facilitated through new policy initiatives, incentives and legislation. This will include intensive networking of capabilities and facilities within the country.

Rigid Quality Standards, and Accreditation of testing and calibration laboratories according to international requirements, will be given an enhanced push to enable Indian industry to avoid non-tariff barriers in global trade.

A comprehensive and well-orchestrated programme relating to education, R&D and training in all aspects of technology management will be launched. To begin with, Indian Institutes of Management (IIMs), Indian Institutes of Technology (IITs) and other selected institutions will be encouraged to initiate these programmes.

Promotion of Innovation

Innovation will be supported in all its aspects. A comprehensive national system of innovation will be created covering science and technology as also legal, financial and other related aspects. There is need to change the ways in which society and economy performs, if innovation has to fructify.

Industry and Scientific R&D

Every effort will be made to achieve synergy between industry and scientific research. Autonomous Technology Transfer Organizations will be created as associate organizations of universities and national laboratories to facilitate transfer of the know-how generated to industry. Increased encouragement will be given, and flexible mechanisms will be evolved to help, scientists and technologists to transfer the know-how generated by them to the industry and be a partner in receiving the financial returns. Industry will be encouraged to financially adopt or support educational and research institutions, fund courses of interest to them, create professional chairs etc. to help direct S&T endeavours towards tangible industrial goals.

There has to be increased investments by industry in R&D in its own interest to achieve global competitiveness to be efficient and relevant. Efforts by industry to carry out R&D, either in-house or through outsourcing, will be supported by fiscal and other measures. To increase their investments in R&D, innovative mechanisms will be evolved.

Indigenous Resources and Traditional Knowledge

Indigenous knowledge, based on our long and rich tradition, would be further developed and harnessed for the purpose of wealth and employment generation. Innovative systems to document, protect, evaluate and to learn from India's rich heritage of traditional knowledge of the natural resources of land, water and biodiversity will be strengthened and enlarged. Development of technologies that add value to India's indigenous resources and which provide holistic and optimal solutions that are suited to Indian social-cultural-economic ethos will be developed. A concerted plan to intensify research on traditional systems of medicine, so as to contribute to fundamental advances in health care, and leading to commercialisation of effective products will be undertaken; appropriate norms of validation and standardization will be enforced. A purposeful programme to enhance the Indian share of the global herbal product market will be initiated.

Technologies for Mitigation and Management of Natural Hazards

Science and technology has an important role in any general strategy to address the problems of mitigation and management of the impacts of natural hazards. A concerted action plan to enhance predictive capabilities and preparedness for meeting emergencies arising from floods, cyclones, earthquakes, drought, landslides and avalanches will be drawn up. Measures will be undertaken to promote research on natural phenomena that lead to disasters and human activities that aggravate them. This will be with a view to developing practical technological solutions for pre-disaster preparedness, and mitigation and management of post-disaster situations.

Generation and Management of Intellectual Property

Intellectual Property Rights (IPR), have to be viewed, not as a self-contained and distinct domain, but rather as an effective policy instrument that would be relevant to wide ranging socio-economic, technological and political concepts. The generation and fullest protection of competitive intellectual property from Indian R&D programmes will be encouraged and promoted.

The process of globalisation is leading to situations where the collective knowledge of societies normally used for common good is converted to proprietary knowledge for commercial profit of a few. Action will be taken to protect our indigenous knowledge systems, primarily through national policies, supplemented by supportive international action. For this purpose, IPR systems which specially protect scientific discoveries and technological innovations arising out of such traditional knowledge will be designed and effectively implemented.

Our legislation with regard to Patents, Copyrights and other forms of Intellectual Property will ensure that maximum incentives are provided for individual inventors, and to our scientific and technological community, to undertake large scale and rapid commercialization, at home and abroad.

The development of skills and competence to manage IPR and leveraging its influence will be given a major thrust. This is an area calling for significant technological insights and legal expertise and will be handled differently from the present, and with high priority.

Public Awareness of Science and Technology

There is growing need to enhance public awareness of the importance of science and technology in everyday life, and the directions where science and technology is taking us. People must be able to consider the implications of emerging science and technology options in areas which impinge directly upon their lives, including the ethical and moral, legal, social and economic aspects. In recent years, advances in biotechnology and information technology have dramatically increased public interest in technology options in wide ranging areas. Scientific work and policies arising from these have to be highly transparent and widely understood.

Support for wide dissemination of scientific knowledge, through the support of science museums, planetaria, botanical gardens and the like, will be enhanced. Every effort will be made to convey to the young the excitement in scientific and technological advances and to instill scientific temper in the population at large. Special support will be provided for programmes that seek to popularize and promote science and technology in all parts of the country. Programmes will also be developed to promote learning and dissemination of science through the various national languages, to enable effective science communication at all levels.

A closer interaction of those involved in the natural sciences and technology, social sciences, humanities and other scholarly pursuits will be facilitated to bring about mutual reinforcement, added value and impact.

International Science and Technology Cooperation

Scientific research and technology development can benefit greatly by international cooperation and collaboration. Common goals can be effectively addressed by pooling both material and intellectual resources. International collaborative programmes, especially those contributing directly to our scientific development and security objectives, will be encouraged between academic institutions and national laboratories in India and their counterparts in all parts of the world, including participation in mega science projects as equal partners. Special emphasis will be placed on collaborations with other developing countries, and particularly neighbouring countries, with whom India shares many common problems. International collaboration in science and technology would be fully used to further national interests as an important component of foreign policy initiatives.

Fiscal Measures

Innovative fiscal measures are critical to ensure successful implementation of the policy objectives. New methods are required for incentivising R&D activities, particularly in industry. New strategies have to be formulated for attracting higher levels of public and private investments in scientific and technological development. A series of both tax and non-tax fiscal instruments have to be evolved to ensure a leap-frogging process of development. The formulation of a focused strategy and the designing of new methods and instruments requires inputs from economists, financial experts and management experts and scientists. For this purpose, the apex S&T advisory body will constitute a dedicated task force to suggest appropriate fiscal measures to subserve the policy objectives.

Monitoring

Effective, expeditious, transparent and science-based monitoring and reviewing mechanisms will be significantly strengthened, and wherever not available will be put in place. It will be ensured that the scientific community is involved in, and responsible for, smooth and speedy implementation.

The New Vision

To build a new and resurgent India that continues to maintain its strong democratic and spiritual traditions, that remains secure not only militarily but also socially and economically, it is important to draw on the many unique civilizational qualities that define the inner strength of India; this has been intrinsically based on an integrated and holistic view of nature and of life. The Science and Technology Policy 2003 will be implemented so as to be in harmony with our world view of the larger human family all around. It will ensure that science and technology truly uplifts the Indian people and indeed all of humanity.

Appendix D

Employment Potential : Address at the International Seminar on Income and Employment Security in India

Dr. A.P.J. Abdul Kalam,

Employment Generation and

I am delighted to participate in the International Seminar on Income and Employment Security in India organized by the Institute for Human Development in collaboration with International Labour Organisation, Planning Commission, National Commission on Informal Sector and Ford Foundation and deliver the inaugural address. My greetings to the organizers, government functionaries, planners, HRD specialists, non-government organisations, trade unions, international organisations and academicians participating in this Seminar. I would like to discuss on the topic 'Employment Generator and Employment Potential'.

Law of Development

I was studying the development patterns and the dynamics of connectivity between nations, especially in trade and business. As you all know the world has few developed countries and many developing countries. What is the dynamics between them and what connects them? Developed country has to market their products in a competitive way to different countries to remain as developed country. The developing country, (if the nation has the vision) to get transformed into developed country; they too have to market their products to other countries in a competitive way. Competitiveness is the scaling factor for both type of nations. Competitiveness has three dimensions: quality of the product, cost effectiveness and supply in time. Indeed this dynamics of competitiveness in marketing of products by developing and developed countries is called the law of development. Competitiveness comes from technology, administration, management, work environment and employee productivity. India is ranked 56 in terms of Growth Competitive Index and 127 in terms of Human Development Index. We have to work together to become 10 in the Growth Competitive Index and 20 in Human Development Index.

Adding Value to Education Through Entrepreneurship

There has been substantial growth in our higher educational system and we are generating over 3 million graduates every year and over seven million plus two level candidates who are aspiring for employment. However our employment generation system is not in a position to absorb all these youth leading to increase in educated unemployed, year after year. This situation will lead to instability in the social structure. We need higher education backed by employment opportunities. A multi-pronged strategy is needed to make education more attractive and simultaneously create employment potential — how do we do that?

Firstly, the educational system should inculcate by adding syllabus of entrepreneurship and prepare the students right from the school and college education to get oriented towards setting up of the enterprises which will provide them creativity, freedom and ability to generate wealth. Apart from entrepreneurship, the youth should have the spirit that "we can do it". Secondly, the banking system should provide venture capital right from every village level to the prospective entrepreneurs for undertaking new enterprises. Banks have to be proactive to support the innovative products for enabling wealth generation by young entrepreneurs. Thirdly, there is a need to identify marketable products and enhancement of purchasing power among the people. This can come through the implementation of mega programmes such as PURA, Interlinking of Rivers, Infrastructural missions. Power missions and Tourism. The universities and schools should become a facilitator for creating this entrepreneurship scheme through the support of the banking system and the marketing system. This will enhance value to the education and create the motivation for the students.

National Employment Status

As per the estimate of the Planning Commission, the total number of people eligible for employment at present is approximately 400 million. Out of this, nine per cent are unemployed which works out to around 36 million. In addition, there is a need to find value added employment for 10 per cent of those employed in the agriculture sector in the rural areas. Our attempt hence should be to find gainful employment for around 76 million people. What are the actions needed to generate employment for 56 million people. This will add to our productivity and will ensure a sustained 10 per cent GDP growth for the decade, which is an essential need for India to become a developed country before 2020.

Employment Generators

Let me share with you how to generate employment. The most important sectors for sustainable national development are Agriculture, Education, Healthcare, Water and Energy. The common thread that will run across these would be the three connectivities of PURA namely physical connectivity, electronic connectivity and knowledge connectivity. These three connectivities will lead to economic connectivity. One of the ways by which the rural agriculturists could increase their earnings is by value adding to the agricultural produce by processing and manufacturing. The farmers, either individually or through their co-operatives would market processed and value added items instead of marketing the raw materials. This increase in the value-addition taking place in the rural area itself is an indicator of the society moving towards prosperity and knowledge era. I would like to suggest eight schemes that can bring large-scale employment and prosperity to our people in the rural sector.

Bio-fuel Generation

We have nearly 63 million hectares of wasteland available in the country, out of which 33 million hectares of wasteland have been allotted for tree plantation. Certain multi-purpose bio-fuel plants can grow well in wasteland with very minimum input. Once cultivated, the crop has fifty years of life. Fruiting can take place in these plants in two years.

Bio-fuel plants grown in parts of the wasteland, for example, 11 million hectares, can yield a revenue of approximately Rs. 20,000 crore a year and provide employment to over 12 million people both for plantation and running of the extraction plants. It will reduce the foreign exchange outflow paid for importing crude oil, the cost of which is continuously rising in the international market. The bio-fuel is carbon mono-oxide emission free. The oil can also be used for soap and in candle industries. De-oiled cake is a raw material for composting and the plantation is also good for honey production. We should absorb best of the technologies available worldwide and start commercial operation immediately. One time investment needed for bio-fuel plantation to production in 11 million hectares will be approximately Rs. 27,000 crore. The capital equipment and investment in plant and machinery can come from bank loans and private sector entrepreneurs. I have seen the progress in bio-fuel plant cultivation, preparation of seedlings, tissue culture and development of non-toxic hybrid varieties in Tamil Nadu Agricultural University, Coimbatore. They have also worked from processing of seeds to bio-fuel production by the indigenous design and development of bio-fuel plants. Anand Agriculture University at, Anand (Gujarat) has also made progress in the bio-fuel cultivation and processing in Gujarat. Bio-fuel plants can be grown in a number of states in the southern, western and central part of the country.

Wasteland Development—opportunities

ICRISAT with its international experience of working in arid regions, have developed short duration, disease and drought resistant varieties of important crops beneficial to our farmers. They have introduced various tillage practices and nutrition management techniques, to boost crop yields even under drought stress. This technology will enable us to reclaim 5 more million hectares of the 33 million hectare of wasteland allotted for productive farming. As per the experts, this will result in the deployment of 15 million people for the dryland cultivation.

Water Harvesting

Water harvesting should become mandatory for all. To improve water table, we need to build check dams, develop water sheds, desilt ponds and rivers, clear the inlcts and outlets to the ponds and water bodies and recharge the wells. If our rural areas are made to have the operational water bodies, recharging of the wells will take place. The task is totally labour intensive and nationwide implementation of this scheme will provide employment for six million persons for more than three or four years. Even after this period many people will be required to maintain water bodies, produce clean water and arrange supply. The scheme will result in increase in storage capacity of water bodies and create additional irrigation potential of the land in the region and enhance agricultural productivity.

Bamboo Mission

This mission envisages an integrated programme of expansion of plantations of Bamboo species, the scientific management, and promotion of community level value addition and entrepreneurship. This will enable presenting the raw material for the industries and the industry to access and apply modern technology for producing globally competitive new generation bamboo products. This includes setting up of clusters of small value addition processing units, near the resource for employment generation and benefit to the local small entrepreneur. Processed raw material suitable for ultimate use in industry / handicraft sector will be required, for economizing handling cost of raw material to the location of industry proposed to be set up in different parts of the country.

The programme envisages the cultivation of bamboo over two million hectares and promotion of technology and networking for enhancing the trade. The economic and social benefit from these activities, will lead to the creation of 8.6 million jobs and the market opportunities worth over Rs. 6,500 crore with an investment of Rs. 2,600 crore. This will be useful for the additional development of the north-eastern region. We should use the management personnel, industrial experts and business houses to provide international market connectivity for our bamboo products leading to sustainable growth, wealth generation and positive contribution to the growth of GDP. The institutions like National Institute of Design, Ahmedabad, Indian Institutes of Management, Industrial houses, State Government and Ministry of Agriculture can work together to generate number of Bamboo enterprises in different rural clusters.

Converting the Fly Ash as Wealth Generator

As you are aware, the use of coal for power generation results in increased quantum of fly ash production, which has reached about 100 million tonnes per year. All out efforts are needed to utilize this fly ash not only from environmental considerations, but also to avoid land usage for fly ash dumping. Though there has been a steady progress in fly ash utilization from 1990, we have a long way to go to reach the target of 100 per cent fly ash utilization. It is reported that the agricultural increase of grains is around 15 per cent, green vegetables 35 per cent and root vegetables 50 per cent, when fly ash is mixed with the soil. Toxicity tests have proved that there is no toxic element due to fly ash. But it has higher nutrients due to increased availability of iron and calcium. The fly ash can become a wealth generator by making use of it for producing "green building" materials, road, agriculture etc. Presently, the fly ash utilization is in the range of 33 million tonnes per year and providing employment for over 50,000 personnel. For full utilization of the generating stock, will provide employment potential for three hundred thousand people and result in a business volume of over Rs. 4,000 crore.

Textile Industry

Textile industry is very important for the Indian economy. The basic raw material is cotton. India is the third largest producer of cotton in the world. However, compared to the world average of 700 kgs of seed cotton per acre we produce only 350 kgs of seed cotton per acre. It is indeed a technological concern for the nation. Some of the industries have adopted a village in Punjab, which has brought out a cooperative movement of the farmers, scientists, trainers and the industry and launched for cultivating cotton in over 1,200 acres. A training programme was launched for farmers starting with soil characterization, matching the cottonseed to soil, water and fertilizer management.

The project resulted in increasing the average seed cotton yield of the village from 450 kg per acre to 950 kg per acre. This led to the seven-fold increase in net return per acre, due to considerable reduction in input costs. It is worth noting that in this case the yield is above the world average. This model has already been replicated in ten villages and can be emulated by many cotton-growing regions of the country, India can definitely produce 25 per cent of the total world production of quality cotton compared to the existing 12 per cent leading to revenue of over Rs. 25,000 crore a year for the nation. The Chairman for Standing Committee on Cotton—Indian Cotton Mills Federation has said this is a feasible proposition and they are working to implement it enthusiastically in partnership with the farmers.

I would like to link cotton production to garment export business, which is a low investment and large volume employment generator. India is presently, exporting six billion dollars worth of garments, whereas with the WTO regime in place, we can increase the production and export of garments to 18 to 20 billion dollars within the next five years. This will enable, generation of employment in general and in rural areas in particular. By tripling the export of apparels, we can add more than 5 million direct jobs and 7 million indirect jobs in allied sector, primarily in the cultivation of cotton. Concerted effort is needed in cotton research, technology generation, transfer of technology, modernization and upgrading of ginning and pressing factories and aggressive marketing strategy.

Healthcare

Another area, which is an employment generator, is the health care industry. We have only one doctor for one thousand eight hundred people, whereas in some of the developed countries the doctor to population ratio is 1:600. For providing quality health care to all of our citizens, we would need at least doubling the strength of doctors and paramedical staff in the whole country. The investment for this, need not necessarily come from government alone. Hospitals can be set up by the private sector with certain tax concessions and subsidized infrastructural support.

Setting up of 30,000 static tele-medicine stations distributed in 30,000 key locations, within the zone of 3 lakh villages and providing 20,000 mobile tele-medicine units will enable reaching of quality health care closer to every home, which are connected to the district, state government hospitals, corporate hospitals, super specialty hospitals in the country. This is possible as India has a network of satellite communication.

How to reach health care for the large number of our population? An innovative method has come into action in certain states. This system provides free health cover to the citizens who are members and pay Rs. 10 per individual per month as an insurance premium. State and Central Government can sponsor this insurance scheme involving payment of a small premium of Rs. 10 per individual per month by the citizens in different states of the country. Such an insurance cover should be able to provide for all types of diseases including expensive open heart surgery. A consortium is required to be formed, in different states between the government, insurance agencies, corporate hospitals and NGO's for providing integrated cost effective health care. The scheme when fully operational, can provide direct employment for additional 6,00,000 doctors and 1.2 million paramedical staff. Apart from providing health care to citizens, these corporate hospitals can attract large number of medical tourists to the country in view of our competitiveness in treating complex diseases. I am very happy to know that the scientific community for health is working on anti-HIV vaccine development, which is in an advanced stage of development. The success of this programme will be a breakthrough in containing the HIV virus.

Village Knowledge Centres

With the kind of awareness and opportunities available in ICT, it will soon become a reality wherein every one of our villages will have computers and connectivity available. These would be the window to the world of knowledge for our villages and also to reap the benefits of our e-governance, tele-education, tele-medicine, e-commerce and e-judiciary initiatives. In spite of the all pervasive nature of the computers they would still be far away from being a truly friendly access device for our villagers. We would need in such cases, a human intermediary who would act as the village information officer. He will be the extended eyes and ears of the villager to the world of knowledge. India has approximately 2.3 lakh Village Panchayats. I visualize establishment of village knowledge centres in these Panchayats to empower the villagers with the knowledge and to act as a nodal centre for knowledge connectivity for the villagers. The knowledge centre from which the villagers would access the information through the village information officer can also be used for collection, digital storage and dissemination of village specific information pertaining to any relevant information to the villagers. This will provide direct quality employment to over one million who will be instrumental in promoting higher level of wealth generation in our rural sector.

The eight schemes discussed so far, have the potential to generate approximately a total of 56 million direct employments during the next 5 years. The other employment avenues in the government and the private sectors based on core competence in areas such as marine products in coastal region, tourism in different states, enhanced employment in ICT sectors including knowledge products and the precision manufacturing sector has a great potential for creating additional productive employment.

Conclusion

For the nation today what is needed is a combination of technology, innovation, leadership and inspired workforce. Let us analyze the dynamics of good organizations for national development. Developed India as defined can be only powered by economic strength.

- The economic strength is powered by competitiveness.
- The competitiveness is powered by knowledge power.
- The knowledge power is powered by technology and innovation.

- The technology and innovation is powered by resource investment.
- The resource investment is powered by revenue and return on investment.
- The revenue is powered by volume and repeat sales through customer loyalty.
- The customer loyalty is powered by quality and value of products.
- Quality and value of products is powered by employee productivity and innovation.
- The Employee Productivity is powered by Employee Loyalty, employee satisfaction and working environment.
- The Working Environment is powered by management stewardship.
- Management stewardship is powered by invisible Leadership.

The invisible leadership means exercising the vision to change the traditional role from the commander to the coach, manager to mentor, from director to delegator and from one who demands respect to one who facilitates self respect. I am sure all the institutions and the future aspiring institutions are spearheaded by invisible leadership. For a prosperous and developed India, the important thrust will be on the growth in the number of Invisible Leaders and innovative organisations which can provide productive and wealth generating employment for all the youth of the nation.

My best wishes to the seminar participants for fruitful discussions on this vital issue leading to the provision of useful recommendations for the mission of providing income and productive employment.

May God bless you.

Index

C

D

E

O

P

Q

R